Kyle Froman

About the Author

In his career as a reporter, ANDREW KIRTZMAN has written a biography of Rudy Giuliani, covered more than a dozen national political campaigns for print and television, and hosted two of New York's most widely watched public affairs shows. He has spent months at a time reporting abroad, covering the fallout from the Iraq war in Israel, Gaza, and the West Bank. In September 1999, *Brill's Content* magazine named Kirtzman one of New York's 10 Most Influential Journalists. In 2003, his week-in-review feature "Kirtzman's Column" won an Emmy Award for outstanding political programming.

BETRAYAL

Also by Andrew Kirtzman

Rudy Giuliani: Emperor of the City

BETRAYAL

The Life and Lies of Bernie Madoff

Andrew Kirtzman

HARPER

NEW YORK . LONDON . TORONTO . SYDNEY

HARPER

A hardcover edition of this book was published in 2009 by HarperCollins Publishers.

BETRAYAL. Copyright © 2009, 2010 by Andrew Kirtzman. All rights reserved. Printed in the United States of America. No part of this book may be used or reproduced in any manner whatsoever without written permission except in the case of brief quotations embodied in critical articles and reviews. For information address HarperCollins Publishers, 10 East 53rd Street, New York, NY 10022.

HarperCollins books may be purchased for educational, business, or sales promotional use. For information please write: Special Markets Department, HarperCollins Publishers, 10 East 53rd Street, New York, NY 10022.

FIRST HARPER PAPERBACK PUBLISHED 2010.

Library of Congress Cataloging-in-Publication Data has been applied for.

ISBN 978-0-06-187077-4

10   11   12   13   14   DIX/RRD   10   9   8   7   6   5   4   3   2   1

It was no accident that I became a journalist. My parents, Marvin Friedman and Doris Kirtzman, have always been passionate about news and newspapers, and more perceptive about the world than any two people I know. They've inspired me so many times and in so many ways, and I appreciate them more each day. This book is dedicated to them, with my love and gratitude.

Contents

BETRAYAL

Introduction

The old man was thrilled. Rolls-Royces and Jaguars were pulling up to the valet stop outside of Palm Beach's Club Colette. Under the warm night sky, guests in black tie exited their cars and made their way into Carl Shapiro's ninety-fifth birthday party.

They entered a large white tent filled with gorgeous exotic flowers that scented the air as the help passed around flutes of champagne. Inside the restaurant, waiters carrying trays of Iranian caviar and blinis stood ready as guests made their way to tables overflowing with red and peach roses. Steve Lawrence and Eydie Gorme warmed up in the wings.

The problems of the outside world seemed a distant matter on this February evening in 2008. Millions of Americans outside of Palm Beach were coming home from work anxious about the future. The papers were filled with warnings of dark times. Federal Reserve Chairman Ben Bernanke testified in Washington that the economy was deteriorating. The foreclosure rate on subprime mortgages was ballooning. The stock market continued its four-month, 2,000-point slide.

Yet the world of Club Colette seemed perfect tonight.

Carl's three daughters, Ronny, Ellen, and Linda, had sweated every detail to honor their father, a patriarch of American philanthropy whose name graced a legion of hospitals, university buildings, and research facilities. Guests speculated that the party had cost Carl's family seven figures. The next day's *Palm Beach Daily News* would gush that the club had been "espaliered with enough greenery and orchids to make Marie Antoinette weep with envy."

The party was a decidedly Jewish affair. Credit card mogul Howard Kessler walked into the room not long after presenting a private performance of the Israeli Philharmonic string quartet in his living room. Financier Abe Gosman had recently lost half a billion dollars from the collapse of his business, but it was just a battle scar in the eyes of this crowd, which welcomed him in his formal wear for another night on the circuit. Shapiro's son-in-law, Bob Jaffe, fabulously wealthy by marriage, impeccably attired and deeply tanned, mingled with the crowd. Carl took his place at the head table, reserved for his family and a few close friends. Bernie and Ruth Madoff were as close as they came, and they joined the group. Bernie sidled up right next to his old mentor.

It had been almost fifty years since Shapiro, founder of the women's clothing manufacturer Kay Windsor, had discovered Madoff, fresh out of school and chomping at the bit to conquer Wall Street. He'd given Bernie his big break, throwing him some money to invest as a kind of test, and he had performed brilliantly. A half-century later, Madoff was a titan of Wall Street, and he had more than returned the favor. He'd invested Shapiro's fortune and made him a multimillionaire.

Carl was a slight, gray-haired eminence, startlingly spry for a man whose one-hundredth birthday wasn't far off the horizon. Bernie was almost 70 but in the prime of his career. His hair had grayed, but he wore it long, bushy, and combed back, an elegant look for a mogul who fussed over every detail of his appearance.

They chatted about small things, mostly family stories mixed with updates on their golf games, the sort of things old lions talk about when they get together in public for dinners and charity events. They had an ease with one another that had developed over the decades, back to the days when they were just Carl the garmento and Bernie the stockbroker. All the years and the stratospheric wealth they'd accumulated since then seemed to fade away.

It was a glorious night. A cake the size of a dining-room table was presented to Shapiro. New England Patriots' owner Robert Kraft presented him with a jersey emblazoned with the number 95. Carl took the microphone and sang "I've Grown Accustomed to Her Face" to his wife, Ruth. Steve and Eydie serenaded them with their favorite song, "It Had to Be You." The 95-year-old glowed with the spirit of a young man still in love with his girl.

Bernie, as usual, kept a low profile. Between courses, the room buzzed with the patter of socialites flitting from one table to the next, but he sat in his chair, making quiet small talk with the people around him. Except for a dance with his wife, he barely left his seat.

In truth, Madoff hated events like these. People never stopped looking at him when he went out in Palm Beach, even if he shrank into chairs and eluded the spotlight. The founder of Bernard L. Madoff Investment Securities LLC was such a successful investor that people literally begged him to invest their savings for them, sometimes in their black ties at events like these. Always, he'd politely demur. It was Jaffe, Shapiro's son-in-law, sitting a few chairs from Madoff, who took their money on Bernie's behalf, if they were lucky.

For all his reluctance, Madoff was in many ways the real center of attention this evening. For Jewish Palm Beach society, this was a rare sighting of the reclusive money manager. Easily half the men and women in this room had their money in Madoff accounts; many of them had invested every cent of their savings with him. It was a status symbol in this crowd to be accepted as his client.

Shapiro was a figure of respect in this circle, but Madoff was the source of their joy. Some made it a point to toast his name at their private dinner parties. Some would approach him in public and show their appreciation with tears in their eyes. People looked at Bernie Madoff with awe. He had, after all, made them rich.

Some could attend events like these only because of Madoff's success with their money. It was Madoff who freed them to dance into the night with barely a worry about the economic collapse looming outside of their beautiful world. Madoff, not Shapiro, was the real father figure in this room.

What could he have been thinking, sitting at his table observing this river of the wealthy? This gathering of the ruling class comprised a club he had longed to join since the age of 13. They were accomplished men and women, giants of industry who had scaled the heights of American business. Madoff wasn't as bright as they, a fact he'd known from an early age. He had achieved his success as an outsider, the architect of an unglamorous, fringe business he had built outside the gleaming walls of the New York Stock Exchange, and for years the barons of Wall Street sneered. Now he was one of them.

What did it feel like to realize that no one there, not even Carl Shapiro, knew the truth? All those social climbers in black tie whose futures he held in his hands didn't seem to have a care in the world tonight as they bathed in their self-satisfaction. How could they know that their fortunes didn't exist, and never had? How could they know that, instead of Bernie fueling their extravagant lifestyle, they had been fueling his? How could the women know that before the year was out, they'd be lining up at pawn shops, desperate to sell the jewelry that glittered around their necks tonight? Or that they and their husbands would be forced to sell their mansions or luxury condos?

How could Carl Shapiro know that the man he loved as a son had stolen hundreds of millions of dollars from him to keep his scam afloat? How could he know that in ten months,

Bernie Madoff would confess that his remarkable success was all a lie? How could they know this glorious evening that this revered man was betraying them all?

On this magical night, as Carl and his wife danced to Steve and Eydie's serenade, Bernie Madoff sat in his chair and wore the look of a decent man.

Thirteen months later, Madoff arrived at a Manhattan courthouse wearing a bulletproof vest under his charcoal suit.

Police ushered him past a roped-off mass of reporters, television cameramen, and newspaper photographers who had traveled from around the world to record the moment. Plainclothes police officers escorted him upstairs to a large courtroom filled with dozens of former clients who had been invited by the court to face him at close range. Over a dozen U.S. marshals formed a human chain in back of the courtroom, each of them scanning the room for signs of potential trouble.

Madoff stood at the defendant's table staring straight ahead at Judge Denny Chin's empty chair, seemingly determined not to turn around and risk eye contact with one of his former investors. Finally, Judge Chin entered the room and read the charges against him. Madoff fidgeted nervously as he uttered his first public words since his arrest on December 11.

"Guilty," he said.

He pulled out a written statement and began reading it in a voice so soft the judge had to ask him to speak up.

Your Honor, for many years up until my arrest on December 11, 2008, I operated a Ponzi scheme . . . for which I am so deeply sorry and ashamed.

I knew what I was doing was wrong, indeed criminal. When I began the Ponzi scheme I believed it would end shortly and I would be able to extricate myself and my clients from the scheme. However, this proved difficult, and ultimately impossible, and as the years went by I realized that my arrest and this day would inevitably come.

I cannot adequately express how sorry I am for what I have done.

And then Bernie Madoff, so widely trusted by his legions of clients that they called him "the Jewish T-bill," was led to a back door and swallowed up by the prison system forever.

His friends back in Palm Beach were stunned by the amount of money involved in his scam. Far from being just a financial guru in their clubby little world, Madoff confessed to running a scheme that ripped off investors around the globe—the largest financial crime in history. His victims ranged from Steven Spielberg and holocaust survivor Elie Wiesel to banks in Switzerland, Austria, Italy, and Spain.

Thousands of clients lost money to Bernie Madoff. Investors lured in by his ability to produce consistent double-digit returns refinanced their homes and poured the cash into their Madoff accounts, only to wake up penniless, and in some cases homeless. Fellow investment advisors, so dazzled by his success that they invested every cent with him, were wiped clean.

Some of Madoff's victims committed suicide. On December 23, 2008, a European investor slit his wrists after learning that he had lost his family's fortune and his clients' investments. Two months later, a British army major shot himself in the head after losing his life savings.

In his telling, Madoff was a good man who got into trouble late in his career and couldn't get out of it. It was an explanation shared by many of his friends, even those he deceived and bankrupted: he was a good man who did a bad thing. But the storyline was as much a fraud as the financial statements he'd manufactured; the facts point to his launching his criminal operation when he was in his twenties.

Many of Madoff's victims had loved him, yet he showed no anxiety, guilt, or remorse as he bankrupted his best friends and even his own sister. Never once did he try to warn them off from his swindle. On the contrary, he successfully hit up Carl Shapiro for hundreds of millions of dollars in the days before his scheme fell apart.

Why did he betray almost every person who ever cared for him? In the course of researching this book, I interviewed more than a hundred people from his past, from the first girl he ever kissed to family members who played in his house as children, the employees at his company, and the inner circle of friends who became his family as an adult. I've read through over two thousand pages of court documents; census, military, marriage and immigration records; congressional testimony; private emails; phone conversation transcripts; and old yearbooks. There is not a hint of criminality or madness in his background.

Yet Bernie Madoff bore scars that few knew about. His parents ran afoul of the law with a shadowy stock trading operation run out of the Madoff home, an eerie precursor to his crimes. Time and again as a kid, he was spurned and humiliated for what was perceived to be his inferior intellect. As his friends excelled in school, he fell hopelessly behind. He was rejected by one girl after another for his shortcomings and relegated to lesser classes and lesser schools. He was assigned to the heap of mediocrity at a young age.

But he excelled at making money, and with it came the stature that once had eluded him. When he couldn't generate as much money as he wanted or needed, he simply invented it.

When America met Bernie Madoff, the economy was tanking. The housing bubble had burst spectacularly, and Wall Street institutions were collapsing from bets on financial instruments that proved worthless. As Americans were growing suspicious that their economy was based on a mirage, his arrest seemed to prove that it was true. He was everyone's worst nightmare.

Madoff's case was a frightening example of nobody asking questions when the going was good. Just as only a few quiet voices questioned the sanity of subprime mortgage derivatives, only a tiny handful of skeptics raised their voices about Madoff's strangely consistent profits, even though no one could replicate his formula or understand his methods. Many on Wall Street suspected he was cheating but did nothing about it. Government regulators proved hapless.

By the night of Carl Shapiro's birthday celebration, Madoff was a warped man. He had created much of the wealth in the room at Club Colette that evening, then yanked it all away in an epic betrayal. The people in their tuxedos and gowns were thrown into a panic, like passengers on the *Titanic*.

This is the story of Bernie Madoff's descent into evil, and the havoc that he wreaked along the way.

1

The Struggler

In the early 1950s, the housewives of Laurelton, Queens would pause their phone conversations every time an airplane roared overhead, so close was their neighborhood to Idlewild International Airport. People would swear that the planes flew so close to their windows they could see the passengers in their seats.

Laurelton wasn't the wealthiest community in New York, but it was idyllic in many ways. The Jews who filled the modest one-family homes had made it out of the tenements of the Lower East Side, Brooklyn, and the Bronx and were luxuriating in the first real houses they had ever known.

It was the smallest of worlds. The boys at P.S. 156 played stickball on its asphalt playground as girls jumped rope. Down the street, the Laurelton Jewish Center hosted Ping-Pong and basketball tournaments for kids after school. Lil

Ed's luncheonette played Rosemary Clooney and Bill Haley and the Comets on its jukebox, and teenagers drank shakes at the counter. After catching a movie at the Itch, as the movie theater was known, the teens would dance in their parents' basements, and steal a kiss or two if they felt brave enough. It was a *Happy Days* existence, the New York Jewish version.

It would prove to be a fleeting moment in time, as Laurelton was just an economic pit stop on the way to the suburban prosperity of Long Island. In the early 1960s, the residents would flee en masse as black families began moving into this homogenous community and the illusion of an all-white Jewish world was fatally punctured.

But for a short period, Laurelton was a treasured home for a community experiencing its first taste of the American dream. People kept their doors unlocked all day, and everything most kids knew of the world could be found on Merrick Road, where you could pick up some egg foo yong at the House of Chang or a charlotte russe at the Four Star Bakery. Times Square was about fifteen miles to the west but a universe away. If the older kids got ambitious, they'd get on the Q5 bus and take it to the neighborhood of Jamaica, twenty minutes away, where they would expand their horizons at Macy's and Gertz.

In the summer of 1953, as lines of moviegoers snaked down Merrick Road to see *From Here to Eternity* at the Itch, an 18-year-old freshman at Ohio State University named Stephen Richards prepared for his summer break back in Queens.

He was a rare breed, an affluent kid from Queens. The

borough was a workingman's haven, separated from Manhattan by two bridges, one tunnel, and an entire way of life. Forest Hills, Stephen's neighborhood, was perhaps the toniest of a pretty unglamorous bunch.

Stephen was a slim young man with the swagger of someone who'd been born to comfort. His father headed up the legal department of the Maryland Casualty Company, one of America's largest insurance firms in the 1950s. At times in his life, Stephen's family had a car and a driver, an almost inconceivable amenity in middle-class Queens. The money allowed him to go to an out-of-state university instead of a local public college, a rare privilege.

He was a frat boy at Ohio State and prided himself on being a sharp New Yorker in a world of midwestern hicks. His roommate, Bob Roman, was an Ohio native and a little sheltered from the rest of the country. So Stephen decided to take him home for vacation and show him the big city. Along the way, a frat brother back in Columbus decided to fix Bob up on a blind date. The frat brother was from Laurelton and knew a pretty 16-year-old blonde girl back home named Joan Alpern.

The adventure led Stephen and Bob to a modest two-story house on a pretty tree-lined street. Sitting on the back porch of the Alpern home was Joan, her younger sister, Ruth, and Ruth's boyfriend, Bernie Madoff. The five hit it off immediately; Joan and Bob were instantly smitten with one another, and the group spent much of the languid summer day together.

Bernie was a thin, lanky boy, 15 years old, with a thick mop of brown hair, small, almond-shaped eyes, and a large, rounded nose. There was little that was remarkable about his looks aside from his long eyelashes, which were so striking that the girls who knew him could describe them a half-century later. Ruthie was blonde, pretty, and spunky, far more animated than her boyfriend. Bernie seemed happy to play the introvert to her extrovert. They seemed unusually bonded for such a young couple; they had an ease in their relationship more typical of an old married couple. Stephen had never met Madoff, but he liked him immediately. He had a relaxed, nonthreatening demeanor; friends would sometimes have a hard time getting him to take things seriously. He seemed to be a good listener, not some showboat.

The two chatted for a while. Stephen asked Bernie if he knew what he planned to do over the summer. "I'm pretty set," Madoff said. "I've already got a job. I'm helping a guy install sprinkler systems." It sounded like a dull way to spend a vacation.

In the week that followed, Stephen and his roommate hit the big city, walking the streets of Manhattan, visiting tourist sites, and soaking up the energy of its postwar boom. Yet their week on the town ended back on that same porch in Laurelton, for Bob was drawn to Joan, who would one day become his wife.

With a light summer breeze pushing through this little porch in Queens, Joan and Bob pursued their burgeoning courtship and Stephen and Bernie got to know one another. The two were developing an easy rapport.

"How'd you make out on that sprinkler job?" Stephen asked.

Madoff smiled. "I installed one sprinkler with the guy and realized I could do this thing myself," he said. "I went into business the next day." He was boasting, a wise-ass showing the college kid how well he was doing. But there was a good deal of substance behind Madoff's cockiness. While his friends were playing basketball or goofing off at the Laurelton Jewish Center, Bernie was on his hands and knees on his neighbors' lawns every day installing sprinkler pipes. It was unglamorous work, digging holes into hardened soil and sinking pipes into the ground. Each long pipe had to snake along a ditch to a central water supply, where a clock and a meter would need to be installed to get the water spraying automatically. It was blue-collar stuff.

Madoff's friends took notice. Donny Rosenzweig would stare out the window of his house, watching Bernie on his knees in the dirt, digging into the Rosenzweigs' grass with a hand spade. Donny was surprised that a friend of his was doing the work of a laborer.

Bernie was the image of a driven young man. Later that summer, he boasted to Stephen that he'd made several thousand dollars from his fledgling sprinkler business, a head-turning sum of money for a kid to earn. Stephen was dazzled. Fifteen-year-old Bernie Madoff had made more money in one summer than Stephen would make in his first year out of college. His destiny as a businessman was obvious; he radiated an entrepreneur's spirit. "This kid is something else," Stephen told people. "He's going to go places."

The forces driving Bernie Madoff's insatiable quest for success were not all good, but that was impossible for Stephen to know at the time. The positive impression Bernie made on him was so deep that it stayed with Stephen for almost a lifetime, coloring decisions he would make for years to come, and ultimately ruining his life.

Years later, Bernie Madoff would tell anyone who would listen that he had grown up poor on the Lower East Side. "I fought my way out of there," he told a guest at his niece's wedding. "I had to scrape and battle and work really hard."

None of it was true.

The Madoff family had followed the classic immigrant path. Bernie's parents, Ralph and Sylvia, were born to Eastern European Jews who had fled to America to escape the terror of government-sanctioned pogroms. Ralph's father, David Solomon Madoff, born in the town of Pshedborsh, Poland, traveled to America in April 1907 with his wife, Rose, making his way from Hamburg, Germany, aboard the mighty Hamburg-Amerika line passenger steamship *Graf Waldersee*. They arrived at Ellis Island thirteen days later and settled in Scranton, Pennsylvania. Both had grown up speaking Yiddish, but David soon learned to read and write English. They would move to the Bronx a few years later, supported by David's salary as a tailor.

Sylvia's parents, Harry and Dinah Muntner, were immigrants as well—he from Poland, she from Romania. They found success in business on Manhattan's Lower East Side, the teeming epicenter of American Jewish immigrant life. Amid the aging tenement buildings, a forest of clotheslines, and the cacophony of a thousand street peddlers, Harry Muntner owned and ran a Turkish bathhouse at 9 Essex Street. It was a major neighborhood gathering spot, where immigrants living in cold-water flats would come for a *shvitz*, or sweat bath, and some gossip inside the steam room, where they'd sit with towels wrapped around their heads in intense heat. Many spoke only Yiddish.

Harry listed the value of the bathhouse at $27,000 in 1930, equal to about $350,000 today. Compared to his downtrodden patrons on the Lower East Side, he was a man of means. His family's housing was secured, for they lived above the bathhouse.

Sylvia Muntner married Ralph Madoff in 1932, at the height of the Great Depression. Ralph worked as an assistant manager at a wholesale jewelry store and moved on to a succession of jobs, but on the marriage license he listed his profession as "credit," a nebulous term if ever there was one. On the line requesting the name of her profession, Sylvia wrote "none." In truth, the two were preparing to plunge into the stock market, a venture that would not end well.

The couple made it out of the Bronx and the Essex Street bathhouse building and moved to Belmont Avenue in Brooklyn and then, in April 1946, to Laurelton, leaving behind the world of tenements and grimy streets for a better life. The home on 228th Street that they bought for $16,000 was a narrow, two-story red-brick matchbox of a house, with a small garage in back and a chimney on the roof. It was nothing particularly fancy, but it was a far cry from the swarming streets outside of Harry's bathhouse.

The Madoffs' new neighborhood was impossibly pretty, with trees six stories high towering over prim houses with small garages and postage stamp-size lawns. They were not glamorous houses, but the homeowners on Ralph and Sylvia's block were proud that they stood alone, not in attached rows as on some of the less affluent streets on the other side of Merrick Road.

Their son, Bernard Lawrence Madoff, was born on April 29, 1938, the year Joe Louis knocked out Max Schmeling at Yankee Stadium and *Time* magazine named Adolf Hitler its Man of the Year. Bernie was named after his father's uncle, a Polish immigrant, and was the middle child, separated by nearly four years from his older sister, Sondra, and by seven years from his baby brother, Peter.

In many ways Bernie was a typical Laurelton kid, playing stickball at the schoolyard and chasing his family dog through the streets. (The pet was a little neurotic, a car chaser, and the Madoffs spent a lot of time frantically running after him.) When he wasn't working, he was riding his bicycle through Laurelton's tree-lined streets or having a burger with his friends at Lil Ed's on Saturday night. The Madoffs seemed to relatives and friends a warm, happy family, though there

were problems in that house that Bernie would all but duplicate in his life years later.

Overall, his was a safe, comfortable, middle-class existence, far from the hardscrabble Lower East Side upbringing
he would portray in his later years. Madoff had deep roots
in the Lower East Side and spent time in his grandparents'
home. But Harry Muntner was no street peddler; he was a
member of the neighborhood's gentry. The myth Madoff had
created of a rags-to-riches upbringing would become integral
to his image. It was one of the many tales he would spin in
the coming years.

Bernie entered P.S. 156 in 1946, shortly before the Commu-
nists seized control in his family's native Poland and the Cold
War began. His five-minute walk each morning took him
past dozens of wood shingle, stucco, and red-brick houses, a
mish-mash of architectural styles typical of Queens' residential neighborhoods.

He'd walk past the giant Laurelton Jewish Center, hub
of the social life of Laurelton's kids, which stood across from
the giant, U-shaped elementary school. They were the twin
monuments of a child's life in Laurelton. When students left
school at 3 p.m., they would walk a few steps down the block
and spend the rest of their afternoon at the Jewish Center's
youth wing.

It was a protected existence. Kids would play Catch a Fly
Is Up against a handball court wall in the school playground
until the teacher blew a whistle, summoning them back to

class. Teachers collected a dollar a week from each student for their bank accounts, and distributed their bank books back to them after each deposit. At lunchtime, many kids walked home, where their mothers waited with white bread sandwiches. Many of Madoff's classmates would recall the era as one of the sweetest of their lives.

Bernie didn't do much to stand out. He was a fairly popular kid, but he was an average to poor student—a huge negative in the world of Laurelton, where the children were expected to live out their parents' single-minded drive for upward mobility. Bernie could shoot a hoop and swim a decent lap, but when it came to academics, he was a star at nothing.

In the course of their handball games and pickup basketball play, Bernie and his best friend, Elliot Olin, hatched a plan to form a gang called the Ravens. They let a few of their friends into this newly elite society, basically a clean-cut group of teenage boys bound by their interest in being cool. They would strut through the chaotic hallways of P.S. 156 together, wearing Ravens sweaters. When a Raven landed a girlfriend, she won the right to wear his sweater, which became a big deal among the girls.

Elliot was the undisputed star of the class. With his curly blond hair, winning smile, and straight-A grades, he was a charismatic presence at P.S. 156. Bernie was his loyal sidekick, and their friendship was deep. But that friendship drew perpetual comparisons between the two, with Bernie usually coming out on the losing side.

Elsa Levine (known as Elsa Lipson then) knew Bernie

Madoff had a crush on her as early as fifth grade. She would catch him stealing glances at her in class. Often he would come to her house looking for her. His shy courtship of her persisted for months. She never agreed to go out with him, though. Bernie was infatuated with her, but she was more interested in Elliot. "He was smarter," said Levine. "He just was cuter and smarter. I don't remember Bernie being that bright at all."

Elsa got what she wanted, rejecting Bernie for his best friend. It was a rebuff that Madoff would remember. Decades later, he ran into Levine at a bar mitzvah and introduced her to his kids. "This could have been your mother," he said.

But it was just one rebuff of many. A year after being turned down by Levine, Bernie asked Marcia Mendelsohn, a girl in his class at P.S. 156, to wear his sweater. Marcia was a brassy, outspoken young girl with blonde hair and a beautiful figure. She was intrigued by this boy with great, thick hair and impressively straight posture. Mostly, though, she appreciated Bernie because he was a gentleman. He was respectful of her, not a wise guy. She accepted the sweater.

For the better part of a year, young Bernie spent much of his free time at Marcia's house, sitting beside her at the piano. She was a gifted musician, and Bernie would while away the afternoons squeezed against her on the bench, watching her play Moonlight Sonata or the Minute Waltz. He was captivated; the high point of his day seemed to be his visit to Marcia's house. He went out of his way to pick her up after lunch and walk her back to school. On weekends, they'd have sundaes at Raabs or Teddy's, or take the Q5 to Jamaica to catch a movie at the Valencia.

For months, Bernie, Marcia, Elliot, and his new girlfriend, Elsa, double-dated constantly. The four were inseparable, strolling along Merrick Road together after school or dancing to records at one of their homes. It would have been perfect but for the fact that one of the four couldn't keep up with the rest academically. Three of them were honors students, who took special classes together throughout the day. Bernie was stuck with the average kids.

At report card time, the cafeteria of P.S. 156 buzzed with the noise of students comparing their grades. The children of Laurelton were expected to excel; there were parents who would fly into rages if their children came home with mediocre results. Munching on their half-chickens and stewed carrots, Elliot, Marcia, and Elsa chattered about their accomplishments with pride. Bernie sat quietly through the excruciating ritual. He wasn't coming close, and his embarrassment was plain to see. "I remember he was always disappointed getting his grades because he didn't make the honor society," Marcia Mendelsohn said. "He wanted to be in the honor classes. We all knew. He was always the struggler."

As time went on, he seemed to grow angry at himself. It was humiliating not to measure up to his best friends. "He had an inferiority complex," Mendelsohn said. "He never felt he was good enough." The chasm between Bernie and his friends grew wider as they continued to move in different, smarter circles. Ultimately, his shortcomings cost him his greatest source of joy. Marcia ended their relationship. "I didn't think he was smart enough," she said. She handed him back his Ravens sweater. She tried to remain friends, but he never returned another phone call from her.

Madoff was living in the shadow of his best friend and being rejected by girls who deemed him mediocre. The rebuffs for not being smart enough came at a vulnerable moment for him, as he was just entering high school. And they continued into his days at Far Rockaway High.

It took a good long ride on the Long Island Railroad to get to the high school from Laurelton, but most parents preferred it over the other school in their district, Andrew Jackson High in St. Albans. Jackson was closer, but both the school and its neighborhood were rapidly integrating, and word spread quickly among parents in Laurelton that it wasn't safe. Far Rockaway High School was about as homogeneous as the community the Laurelton kids were leaving each morning.

The commute to Far Rockaway was initially a communal headache for Laurelton's kids. The track on the Laurelton side was two flights up and outdoors, with a sweeping view of the community's vast sea of small homes. There was scant protection from the elements; rain and snow would splatter them as they waited for their train. The ride took a numbing forty minutes, but after a while the students started to enjoy the routine and turned the commute into a social event. The Wavecrest station at the other end of the ride was close to the beautiful Rockaway beaches; students would hang out on the steps overlooking a roaring ocean. Sometimes they'd head over to Hal's Luncheonette and dance The Fish in front of its jukebox. Students would meet their dates here.

Madoff fit in well. He joined the swim team, where he was a valuable, though not starring, member of the medley relay, able to swim backstroke, butterfly, and freestyle. He swam well enough to earn a lifeguard's job at the beach when he wasn't laying sprinkler systems back in Laurelton. The money from both jobs was swelling the bank account of this driven young man.

Bernie's coach bestowed on him the honor of serving as a locker-room guard, in which capacity he broke up fights between the jocks and sometimes mixed it up with them to show them he was boss. In the mornings, he'd shake his head talking to his friends about how dumb the athletes were and show off his bruised knuckles. "I'm not going to let those idiots think I'm chicken," he told them.

He was a good-natured kid, not given to mean or selfish behavior. There were no episodes of cheating on exams, manipulating his friends, or stealing food from the cafeteria. "He was a happy-go-lucky guy," according to Fletcher Eberle, co-captain of the swim team. "He didn't take things that seriously."

One morning in sophomore English class, students were expected to deliver oral reports on the book they'd been reading for several weeks. Bernie Madoff hadn't even opened one up. His Ravens pal Jay Portnoy was prepared, though. Bernie picked up Portnoy's book and scrutinized it. "Boring," he concluded. "Hardly any pictures." When class began, Bernie was called on to deliver his report. He had no intention of coming clean. "My book is called *Hunting and Fishing*, by Peter Gunn," he announced, triggering snickers from his friends. Peter Gunn was a private eye character on television. He went on to fabricate names, dates, themes, and conclusions. Asked to produce the actual book, he claimed that he had already returned it to the library.

Madoff had the teacher fooled. "Bernie glided through his book report," Portnoy said. His friends protected him. "No one really wanted to see Bernie fry," added. "Nobody could really get mad at Bernie." After class they all congratulated him.

As at elementary school, Madoff was a popular kid. On the train back and forth to Laurelton each day, the "social pecking order was determined by LIRR seating," Portnoy said. "Bernie and Elliot with two of their close neighbors were usually the core group. Others tried to sit near them. If one of the core four was absent . . . others, including myself, would try to sit near them."

But once again, Bernie's best friend stole all the attention at Far Rockaway High. "Elliot was more visible," said a high school classmate. "His relationships with people were more profound. He had charisma. Bernie was a nice kid, with a nice personality. Did he feel inferior to Elliot? I'd say yes."

"When he was in school," this classmate said of Madoff, "he wasn't somebody. He was just another kid. He wasn't that smart."

Meanwhile, Bernie's kid brother, Peter, was showing great academic potential in grade school. His reputation as one of Laurelton's brightest would be borne out in later years, when he won admission to Brooklyn Tech, a prestigious high school, while Bernie had plodded along at Far Rockaway High. The comparisons at home, as in school, must have hurt.

The irony is that all the classmates who judged him and all the girls who rejected him were wrong about Bernie Madoff. As he grew older, he would prove to be plenty smart. He wasn't book smart, and he showed no signs of a superior intellect. But he had a gift they did not, and he would become far more successful than almost all of them. The gift was his understanding of money.

Money was something Madoff could produce in prodigious quantities. The more he made, the more approval he won, a big ego boost for a kid in desperate need of one.

The naked display of material success is a time-honored aspect of American suburban life. The residents of Laurelton had not yet made it to the Long Island suburbs; they were economic strivers in a community that was a step away. But proving their financial success was, if anything, more important here, as residents feverishly pursued their suburban ambitions.

Irene Shapiro, who was raised in Laurelton in the 1950s, put some of her recollections of Laurelton to paper. She wrote that the parents of Laurelton "had all made it out of Brooklyn, the Bronx, etc., and with this accomplished they began to look around at each other and rate and evaluate and compare. And everything they compared cost money."

"The more money, the better your things," she wrote. "The girls counted cashmere sweaters, Pappagallo shoes. They compared beach clubs. Every rank was well known: El Patio and Sands toward the top, Capri and Silverpoint at the bottom, locker better than cabana, cabana in aisle A better than aisle F." It was a distinctly Jewish, postwar psychology, she recalled:

> It wasn't the world of Talmud or Torah studies. They identified with the Jews that marched into Tel Aviv, not the Jews who were marched into Dachau.
>
> Attendance [at synagogue] was mainly on the High Holy Days, and what you wore on these days was a serious and important choice: The women DRESSED. The men attended services, with seating determined by money. At some point in the day, the name of each member was read aloud from a list—and each member was expected to stand when his name was called and state his yearly donation. It was quite a telling ritual.
>
> The more money, the more the esteem. The more money, the more people looked up to you.

It was a message hammered home to Bernie Madoff and most of the other students at Far Rockaway High. How could it not have made an impression on him? On weekends, girls at their high school would ditch him and his friends for the far more affluent teenagers of Long Island's Five Towns.

Laurelton was a community on the frontier of Gotham, nestled between the worlds of the city and suburbia. For many, it was an accomplishment to make it this far, away from the inner city. But as you traveled farther, crossing from Queens into Nassau County, Long Island, a new level of wealth and prosperity became visible.

Laurelton's Jewish residents called the Five Towns "the gilded ghetto." Only minutes away, they stood as a bastion of success, the next step on the quest for upward mobility. An upscale area some residents described as "suburban chic," the Five Towns held a collection of villages and hamlets: Hewlett, Woodmere, Cedarhurst, Lawrence, and Inwood. Bernie passed through the Five Towns on his way to school each day, getting a constant reminder of the wealth that existed so close to home.

Five Towns residents inspired envy in Laurelton's teenagers and parents alike. The teenagers had nice cars and big allowances and swept the Laurelton girls away to nice restaurants and dance clubs that the Laurelton boys couldn't possibly afford. A date with a Laurelton boy, on the other hand, basically meant a movie and a night spent necking in a parked car.

For Madoff, the lessons he was learning in his personal life and from the world around him were the same: money could be his key to happiness.

• • •

Eventually, Bernie found love at Far Rockaway High.

Ruthie Alpern was three years his junior. She was "sweet, nice, and bright," in the words of Cynthia Arenson, a childhood friend. She dressed smart and pretty, in plaid skirts and button-down shirts. "She wasn't flashy. She was someone who'd dress in a nice wool sweater." She was preppy before the word had been invented. To no one's surprise, her classmates voted her "Josie College" when she graduated.

Ruthie and Bernie got to know each other rushing through the hallways on the way to their classes, which led to burgers at Raabs, movies at the Itch, and walks along the beach near school. Her outgoing personality was a perfect complement to his quieter nature. The two became inseparable. "She would spend most of her life at his house," said Donny Rosenzweig, who had dated Ruthie before she met Bernie. Bernie increasingly became a part of Ruth's family, which, like his, was middle class and Jewish. Ruth's father, Saul, was a bookish accountant who was teaching her the trade in the evenings when she wasn't doing her homework. Years later, Bernie would put Saul's talents to good use.

Bernie had found his biggest fan and future wife. Ruthie saw a spark in him that others did not; she believed in him when others hadn't. They became each other's best friend and closest confidant and stayed joined at the hip for decades to come. It would be a half-century before events would pull them apart.

On June 26, 1956, as President Dwight Eisenhower campaigned for a second term, Bernie Madoff graduated from Far

Rockaway High. The classmates he celebrated with that day were about to fan out to the city's elite public universities and colleges, which offered free tuition to students who earned grades high enough to win admission. But Madoff was heading for the University of Alabama, a strange destination for a Jew from Queens and hardly a prestigious school. He told friends he had been recruited by the university's swim team, an honor. It might have been the case, though kids from the Northeast generally attended the southern university because it was notoriously easy to get into.

The campus was bucolic, a thousand acres of bright magnolias and lazy dogwood trees. It was a school rich in southern tradition, charm, and prejudice, most of which remained unchanged since the days of the Civil War. Six months before Bernie arrived, mobs of students, townspeople, and groups from across the country descended upon the university's home city, Tuscaloosa, trying to stop the school's first black student from attending classes.

It was as far away from Merrick Road as one could imagine, and Madoff stood out among the students passing under the porticos of the university's antebellum houses. Yet the impression he made on the students there was a familiar one. "I didn't think of him as a particularly bright kid," said Madoff's fraternity roommate, Martin Schrager. "There was nothing outstanding about him. He was just a guy in the fraternity house."

At the end of his first semester, Madoff packed his bags before he could even pledge the frat he was living in and said so long to Alabama. He returned home and enrolled at Hofstra College on Long Island, where he'd be far closer to Ruthie.

By the time Bernie returned to Laurelton, people at almost every juncture of his life had judged him as nothing particularly special, and his shortcomings had been made abundantly clear to him. He entered his twenties intent on proving them wrong.

2

The Plan

On a bitter cold night in 1958, the streets of Laurelton lay frozen under a sheet of ice. Jay Portnoy, a college student and former member of the Ravens, stared at the bleak weather from his front parlor window. It was the kind of night to stay indoors.

At about 8 p.m., Bernie Madoff pulled into the driveway of his friend's house. Jay put on his jacket and carefully made his way outside and into Bernie's car. There was a card game, a round of Hearts, planned at a friend's house, and no one was about to let the bad weather get in the way.

Jay got into the front seat with Bernie and the two took off on Francis Lewis Boulevard before making a left onto 135th Avenue. Many of the side streets of Laurelton were so narrow that you could barely drive through the parked cars on both sides of the street. And it was very dark out there. Jay was anxious about the road conditions, but his friend brushed aside his fears. Madoff stepped on the gas and started speeding down the icy road at 40 miles per hour. Jay grew nervous. "Why are you driving so fast?" he asked, grabbing his armrest.

The car barreled down street after street, racing through gauntlets of narrow roads. "Calm down," Bernie said. "Our car's just as wide whether we're going twenty miles an hour or forty miles an hour. It makes no difference." He seemed to be enjoying himself. Jay kept protesting, but Bernie wasn't hearing it. "I'd have just as much of a chance nicking a car if I went slower," he insisted. It was twisted logic. Bernie was risking a crash as he sped through the winter darkness. Jay had no choice but to hold tight all the way to the card game.

Most of Bernie's friends considered him an easygoing guy. But his appetite for risk, and his apparent belief that he was impervious to its consequences, would grow increasingly apparent as his career bloomed. It was a trait he likely picked up at home.

In many ways, the Madoffs were a model Laurelton family. The children were well-behaved and industrious. The parents attended PTA meetings in the evenings and seemed to be commendably engaged in the lives of their kids. Friends and relatives who spent time inside their house on 228th Street saw a happy family.

Ralph Madoff, a big, gruff man with the burly frame of a Teamster, was the dominant force in the family's small, three-bedroom home. He was a bear of a man, the kind you don't mess with, yet sweet and gentle around the kids he was so proud of. His wife, Sylvia, was a more distant figure to Bernie's friends; more often than not it was the Madoff's daughter, Sondra, who seemed to play a maternal role in Bernie's and Peter's lives. On the whole, few if any people saw indications of serious dysfunction.

Years later, Bernie would grow emotional describing his family and fight back tears when he recalled his father's death. But life was a lot more complicated in his household than he ever let on. Illegal activities were taking place inside the Madoff home. Ralph, Sylvia, and apparently Bernie were running a rogue stock trading operation out of their living room.

Friends say Ralph worked as a plumber for several years, but at some point in the 1950s he started to trade stocks. He seemed entranced at first by Wall Street's promise of big, fast money. But years later he would reflect on his experience with bitterness. Riding in a car with William Nasi, who knew him back in Laurelton, Ralph offered him some advice. "Never, ever, ever invest on Wall Street. It's run by crooks and SOBs. I don't trust them. . . . Put your own money in a savings bank and you control it yourself," he said. "A dollar is worth a dollar. Don't let greed get into your psyche."

A friend of Bernie's told Stephen Richards at the time that Bernie, then a student at Hofstra, was working as his stockbroker even though he didn't yet have a securities license. "He's clearing his trades through his father, a stockbroker," Richards's friend explained. Yet there are no records showing that Ralph Madoff ever held a stockbroker's license. It turns out that it was his wife who had two securities companies registered to her name, called Gibraltar Securities and Second Gibraltar Corp. It was unusual, to say the least, for a Laurelton housewife back then to work as a stockbroker. Apparently her husband and her son were trading stocks illegally.

Their activities were shrouded in secrecy. Sylvia failed to file at least one year's worth of financial statements disclosing the company's activities to the Securities and Exchange Commission. That was a breach of SEC regulations, and in 1963 the agency caught wind of it and filed a complaint against her two companies. Rather than face formal hearings, Sylvia admitted that she had failed to disclose their financial condition and shut down the companies.

Financial turmoil was becoming a way of life in the Madoff household. The year Bernie graduated from high school, the government slapped a lien on the family home for almost $9,000. Ralph and three business partners—it wasn't clear what business that was—had defaulted on their withholding taxes. Presumably aware that the IRS would be coming after him, he transferred the deed on the house to his wife soon after the default.

Bernie lived through the ordeals but never talked about them when he grew older. He never told his friends that it was his parents who gave him his passion for the market, or that he learned the trade by working for them. All that is clear is that shortly before they were thrown out of the business, he plunged right into it.

• • • •

On a warm day in June 1960, Bernie graduated from Hofstra with a bachelor's degree in political science, fairly useless for the career path he'd chosen. Director Francis Ford Coppola and actress Lainie Kazan were among the other members of his class, but Bernie had made few friends in the three and a half years he'd spent there after transferring from the University of Alabama.

He had married Ruth Alpern a year earlier, soon after she'd graduated from high school. The wedding ceremony was held inside the grand synagogue of the Laurelton Jewish Center, flanked by walls of stunning stained-glass murals depicting Old Testament scenes. The reception was held one flight down, under a glittering crystal chandelier. It was a joyous union of a couple deeply in love, held in the bosom of Laurelton's Jewish bourgeoisie.

Bernie and Ruth soon moved into their first home together, a one-bedroom apartment in Bayside, Queens, with a rent of $87 per month. They got what they paid for; the building was a charmless six-story redbrick fortress.

Hoping to avoid being drafted, Bernie enlisted in the U.S. Army Reserves. In September 1960, he was commissioned as a second lieutenant and was assigned to the headquarters of the 1st Army, located on Governors Island in New York Harbor. Madoff served for just over three years in what was basically a weekend administrative job, the kind of service that was coveted by young men seeking little pressure and some crucially needed income.

To be judged a success in Laurelton, one usually became either a doctor or a lawyer. Bernie's brother, Peter, would fulfill the high expectations virtually everyone had for him and receive a law degree. Bernie enrolled in Brooklyn Law School in 1960, but it seemed to interest him about as much as his semester at the University of Alabama. He dropped out after a year.

At the age of 22, Madoff began his mission in earnest: Bernard L. Madoff Investment Securities was born. Whatever lessons he'd learned about money and ethics inside the family's house on 228th Street were about to be put to use.

He had set his sights on Wall Street before he'd even finished college. He passed the General Securities Representative exam, qualifying him for a stockbroker's license, the usual ticket for a young grad entering Wall Street. But it was small potatoes compared to the far more difficult General Securities Principal exam Madoff trained for and passed on the same day. As a result, he was licensed not just to work for a brokerage firm, but to run one.

He would claim years later that he started his company with $5,000 saved up from his lifeguard salary and sprinkler system profits. He would say at other times that he got his start with the help of a $50,000 loan from Ruth's father, Saul Alpern. What is clear is that Alpern had high hopes for his son-in-law. He set Bernie up in business and gave him space in his accounting practice on Manhattan's west side. Then he set him free to make his mark.

As the season's crop of fresh-faced young business school

graduates swung their new briefcases through the doors of Wall Street behemoths like Lehman Brothers and Morgan Stanley, Madoff opened the doors to his own firm, a one-man boutique specializing in buying and selling stocks considered too lowly for the big firms to bother with. He didn't have a business school degree, or even a major in finance or economics.

The New York Stock Exchange was pretty much the whole ballgame on Wall Street in those days. For centuries, it had been the prime place for companies to raise money by selling ownership shares to the public, whether to stretch their railroad lines across the continent or open department stores in dozens of cities and suburbs. Except for a small fraction of stocks trading on the American Stock Exchange, the entire U.S. stock market was basically controlled behind the iconic Corinthian columns that marked the Broad Street façade of the Exchange's famous home.

Only new or obscure companies were up for grabs to wholesale traders like Madoff, who were outside of the great exchanges. Their shares were known as penny stocks since that was often their value. Companies like Madoff's traded them "over the counter," in Wall Street parlance, meaning outside of the exchanges and, in Madoff's case, over the counter of his cheap, dusty desk.

The people in Madoff's chosen field were "street-smart, scrappy, tenacious guys," in the words of one of his early competitors, and they fought constant turf wars with one another. They were less refined than the suits working inside the exchanges, outer borough guys more likely to have come from, say, Laurelton than the Five Towns. Madoff had a palooka's Queens accent that emitted a classic *Noo Yawk* sound, which the gentlemen brokers of Wall Street no doubt sniffed at. He was an outsider in every sense of the word, picking up the crumbs the Ivy League traders tossed from the windows of E. F. Hutton and Merrill Lynch and scrounging to make a living from them.

Years before a day trader could open his laptop and check the latest stock prices, over-the-counter brokers had to rely on listings they received each morning. These reports, printed on pink sheets, listed the price the dealer would pay for the shares and the price he would sell them for (the bid and the ask). The dealer pocketed the difference between the two. For smaller over-the-counter stocks, dealers like Madoff could make as much as 50 cents on a dollar's worth of stocks. He'd have to work a lot harder to sell stocks in what were often rinky-dink companies, but hard work wasn't a problem for the former sprinkler salesman.

Working out of Saul Alpern's accounting firm, Alpern and Heller, Madoff began the humbling work of trading penny stocks. He would sit at his desk for hours at the office in the west forties, far from the action downtown on Wall Street, trying to sell off cheap shares of companies too poor or insignificant to merit a listing on the big board. It was an unglamorous life, commuting from a dreary Bayside apartment to an accountant's office in the middle of nowhere each day. But Madoff was conceiving a spectacularly ambitious path to conquer Wall Street at an early age. Working long hours trading stocks for struggling companies, he was eyeing a second, more prestigious arm to his operation: investing money for individuals, preferably wealthy ones. It was the discrete end of the business, and it was often highly lucrative. What he needed was someone willing to trust him with his money.

The angel he sought appeared just a few months later, in the form of a woman's dress maker from Boston.

For a career garmento, Carl Shapiro didn't look the part. He was 47, a slender man with a sleek haircut parted neatly on the side, perfect posture, and a white handkerchief planted in the breast pocket of his finely tailored suits. He was a vision of rectitude, the result of a proper upbringing by a Brookline, Massachusetts, family that was wealthy enough to employ a full-time servant. He had one foot in Boston society and the other in New York's ramshackle garment district, whose streets were perpetually clogged with burly men pushing racks of clothing from one musty building to another. He divided his weeks evenly between the two cities, and his two lives.

He and his father, a pattern maker and Polish immigrant, had founded the Kay Windsor Frocks company in 1939, the year after Bernie Madoff was born. At the time, the garment industry disdained cotton dresses as *shmates*, or rags, to be worn around the house in hot weather. The Shapiros had other ideas. Carl tirelessly evangelized for cotton to a skeptical industry and proved his point by marketing cotton dresses to the masses. His "Private Secretary" line, based on a popular television program of the same name, was "chosen especially for you by Ann Sothern," the show's sultry blonde star. The motto was "The look you love." The dresses were conservative, comfortable, and affordable, priced at $8.95 to $14.95. Twenty years later, Kay Windsor was churning out thousands of dresses a day in twenty-eight factories throughout New England, with annual sales of $22 million. Shapiro anointed himself America's cotton king.

Shapiro met Bernie Madoff in November 1960, as the young man from Queens was kicking off his career in the financial world. "A friend asked me to meet him, maybe throw him a little business," Shapiro recalled. "I had plenty of irons in the fire, so I declined. But my friend insisted."

Attracted to Madoff's street smarts and determination, he gave the 22-year-old a test. Shapiro played the stock market and was earning money in arbitrage, complex stock transactions designed to take advantage of fluctuating share prices in different markets. He assigned Bernie to execute an arbitrage deal. It wasn't the kind of assignment that usually went to untested kids, but Madoff was a confident young man. "In those days, it took three weeks to complete a sale," Shapiro recalled. "This kid stood in front of me and said 'I can do it in three days.' And he did."

When Shapiro looked at Bernie Madoff, he didn't see the scholastic mediocrity with a middling intellect that so many had perceived in Laurelton, Far Rockaway, and Alabama. He saw a prodigy, with an understanding of the financial markets that was sophisticated beyond his years. Like Shapiro, Madoff was a self-made man, a hardscrabble type with a working-class ethic and a Jewish immigrant heritage. The millionaire, who had three daughters but no son, found his protégé.

As a reward for acing the arbitrage assignment, Shapiro handed Madoff $100,000 to invest for him, a potentially career-making pile of cash delivered on a silver platter. Thus began a relationship that would explode into the financial stratosphere over the years, helping to make both men wealthy beyond their wildest dreams. Shapiro would speak adoringly of Bernie Madoff, boasting at points that the two trusted each other so much they didn't have written contracts between them (though that was more sentimental bravado than fact).

In the decades to come, Madoff would befriend Shapiro's family members and do enormously important business with one of them in particular. Bernie and Ruth were so central to Shapiro's life that they sat at the Shapiro family table at events marking Carl's most important milestones, right up to his ninety-fifth birthday party at Club Colette.

Back in the days when the Catskills were the Catskills, it was taken for granted that if you were a Jewish New Yorker of a certain age and income, large chunks of your summer would be spent miles northwest of Manhattan in the Borscht Belt, nestled in the upstate New York capital of Jewish kitsch. Grossingers and the Concord towered over the rest, but in its heyday, there were more than six hundred hotels, bungalows, and summer camps in the sweltering mountain area some called the Jewish Alps. It played host to a generation of Eastern European Jewish families, who cleared out of the Lower East Side in their station wagons each Memorial Day for a summer of suntanning and schmaltz.

On just one grassy road in Woodridge, New York, there were three hotels and four bungalow colonies. The Sunny Oaks Hotel and Cottages was hardly the nicest of them. In the 1940s, a Russian immigrant couple, Chia and Abraham Pendrus, converted an old two-story farmhouse into a four-bedroom, one-bathroom hotel. Over the years, their children and eventually their children's children assumed its management and added bungalows to the property, swelling its summer population to almost a hundred. But it remained a decidedly unluxurious place, its cramped cabanas decorated in cheap wood paneling glued to the walls to hide aging wallpaper.

Yet the Sunny Oaks would live on after so many other Catskills resorts had died. Cynthia Arenson, a third-generation owner, credited the hotel's survival to the private bathrooms her parents had added, a relative luxury among small area hotels. By the early 1970s, its summer residents were all in their seventies and eighties, but they were a loyal bunch; the same elderly couples returned from their Florida homes year after year, all for the privilege of folk dancing in its simple, lodge-style dining hall or sunning around its tiny swimming pool.

Bernie's father-in-law summered at Sunny Oaks every

year. Saul Alpern was "an accountant personified," one resident said, a bookish man who wore horn-rimmed glasses and conservative sport jackets. He wasn't particularly dashing or charming, and he wasn't the type to line-dance. He was a quiet man, content to spend his lazy summer afternoons playing a game of bridge or reading his *New York Times* on his small porch.

The Alperns' cabana consisted of a kitchen, a bedroom, a little deck, and a bathroom so small there was no room for a bathtub. The cabana "was really cruddy," Arenson laughs. "It was very rustic. Not rustic fancy—just rustic."

It was hard to believe that Saul was an affluent man. When he retired, he and his wife bought a place in Florida and continued to summer in the Catskills. They were so frugal that they lugged their dinette set and their large, olive-green refrigerator up to Sunny Oaks when they sold their Laurelton house. But he had retired with money, thanks to Bernie Madoff's explosive success.

Carl Shapiro's money had been a godsend, but Madoff needed, or wanted, more money to grow the investment side of his business, and he designed a concept that would expand it exponentially. He conceived a kind of financial octopus, with tentacles in every direction, each pulling money toward the center. The feeders would draw in the cash, and Madoff would gather it, grow it, keep some of it, and disburse the rest back to the investors. The concept of feeder funds was not unique, but it was a breathtakingly audacious concept for a kid barely out of college. Yet if you had an insatiable appetite for money to manage, it was the perfect system.

The people Madoff turned to as feeders were literally sitting right in front of him each day: his father-in-law and two young accountants in Alpern's office, Frank Avellino and Michael Bienes. None of the three was a licensed securities dealer, and their soliciting of investments for Bernie was illegal. Madoff would contend years later that he had no idea that Avellino and Bienes were unregistered brokers, but it smacked of Captain Renault in *Casablanca* contending that he was "shocked, shocked!" to find gambling going on as he collected his winnings. These were men who worked alongside Madoff.

Like any young man starting out in the world, Bernie looked to his friends, his family members, and their friends for his first business connections. That's where the Sunny Oaks Hotel came in.

By the late 1970s, as one Catskills hotel after another shut its doors or converted into homes for the retarded or infirm, Sunny Oaks was still thriving thanks to its loyal crowd of couples and aging widows. "They played bridge and they folk danced and they line-danced," says Arenson. "They'd have intellectual discussions about Israel, and no one would disagree. It became like an elderly hostel hotel."

Word spread from cottage to cottage that Saul Alpern's son-in-law was making his clients serious money on Wall Street—and the returns were guaranteed. "They'd give you eighteen percent," Arenson recalls. "No more and no less."

One by one, the elderly snowbirds of the Sunny Oaks trekked through the gnat-infested Catskill air to Saul's cabana and asked to be let in on Bernie Madoff's Wall Street money machine. These were not wealthy people, but they had money saved up for their retirement. Some were widows who had inherited a little cash at the death of their husbands. These were Depression people, not the kind to gamble in the stock market, but there was supposedly no risk involved. And Saul was the kind of man you could trust. "My husband didn't want to go into it," Arenson says, "but I said, 'Look, it's Saul Alpern's firm. He's a mild-mannered guy. He's a modest man. He doesn't live high off the hog.' "

More than a dozen residents of Sunny Oaks signed up, handing over $5,000 or $50,000 or whatever part of their retirement account they were willing to hand over. Saul bundled them all into one account and handed it like a gift-wrapped present to Bernie Madoff. Saul didn't invest the money; he was a mere go-between, a tentacle in the Madoff octopus. He was operating way outside the law, but mild-mannered Saul nevertheless went about collecting checks for his son-in-law year after year.

Yet Alpern's success in luring new clients was starting to pale compared with the work of his junior accountants, Avellino and Bienes. According to Bienes, he and Madoff got to know each other after he represented Madoff in a tax audit. Bernie continued to court him when it was over; the two went for nude swims and massages at the New York Athletic Club. At one point, Madoff invited Avellino and Bienes to a party. "I remember my partner, Frank Avellino, and myself and Bernie meeting in the middle of the dance floor," Bienes later told a reporter for *Fortune* magazine. "And we were saying, 'Thanks for having us,' and he said, 'Hey, come on—we're family, aren't we?' And at that moment, he had me. He had me. We were family."

"It really took me," Bienes said, "because he had a presence about him, an aura. He really captivated you."

Thus Alpern's young assistants plunged head-first into the business of soliciting investment dollars for Madoff. As unlicensed brokers, operating in defiance of U.S. securities regulations, they were free to make all sorts of promises that real brokers could not. They weren't printing disclosure statements, submitting government filings, or dealing with the usual restrictions that limit a broker's ability to land a fast deal or pull a quick con. There are no guarantees when you gamble, whether in Las Vegas or on Wall Street, but Frank Avellino and Michael Bienes were promising investors just that: guaranteed returns as high as 20 percent, according to the SEC. When the stock market fell, they promised, the returns would stay the same, along with the investment principal. With a pitch like that, it wasn't hard to find takers. Avellino and Bienes fished for investors among their accounting clients, their friends, and their friends' friends, guaranteeing the impossible.

If the value of the stocks Madoff purchased for his investors fell, someone would have to make up the loss. If it wasn't the customer, who would it be?

No one seemed to question this implausible business proposition. Investors in Madoff's fund were making too much money to care. Most of them, particularly the senior citizens of the Sunny Oaks Hotel, wouldn't live to see the catastrophe that resulted from their decisions. The children who would inherit their Madoff accounts wouldn't be as lucky.

Bernie was on his way. His stock trading operation was burgeoning. In one deal alone, he made $60,000 from the sale of 100,000 shares of a steel company in Corona, Queens. The country's median family income at the time was less than a tenth of that.

He and Ruth were at last able to make the jump to the other side of the middle-class rainbow: suburban Long Island. They purchased a modest ranch-style house in Roslyn, a quintessential middle-class Jewish suburb where schoolchildren played on acres of soft grass, not concrete. She was pregnant with their second son, Andrew, and was caring for his two-year-old brother, Mark. They were a thriving, upwardly mobile family.

In 1970, Stephen Richards and Bob Roman took a flight to Columbus, Ohio, for their tenth anniversary reunion at Ohio State. It had been more than a decade since the two had met Bernie Madoff and his girlfriend Ruth Alpern on her parents' porch in Laurelton. Bob had married her sister, Joan, after courting her that summer. Ruth had since married Bernie. Now Madoff's brother-in-law, Bob had begun to invest with him even before Madoff received his broker's license.

But Stephen had stayed away. "You do it," he said. "We'll watch and see how it goes."

Stephen Richards was on his way to running a small empire of New York City furniture stores and becoming a man of means. Sitting beside his friend ten thousand feet in the air, he started to peruse Bob's résumé at his friend's request. He flipped a page and got an accidental peek at an unrelated document, Bob's monthly statement from Madoff Securities. Richards's eyes widened as he realized that Bernie had made his brother-in-law a wealthy man. "How the hell did you make so much money?" he asked.

"I told you to invest with Bernie ten years ago," Roman replied. It was all the proof Richards needed to sign up as a client of Madoff's. He plunked down $100,000 in a Madoff account when his father died and left Richards an inheritance.

The investment was made through Saul Alpern, for Madoff had handed his father-in-law the entire administrative operation for his investment business and washed his hands of the grubby duty of processing his clients' accounts. Richards mailed Alpern a check with a note: "We've scraped together about a hundred thousand dollars that my father left me and we're looking for this to take care of our sons." He got back a note from Alpern that had been ripped from a small note pad: "Dear Steve and Fran—received your check for a 100k and have mailed it off to the broker. Best regards, Saul Alpern."

About a week later, Saul wrote them again: "Dear Fran

and Steve, I thought you would like the official acknowledgment of your check. Best Wishes, Saul." There were no papers to sign, no disclaimer statements, no notices of investors' rights. Just a small slip of paper torn from a five-dollar receipt book stating "Bernard L. Madoff Securities credited your check . . . $100,000." It was as if he'd bought a pencil sharpener.

Eventually, Alpern wound down his accounting career after the death of his partner, Sherman Heller, and handed his firm over to Avellino and Bienes. The firm had just one client: Bernie Madoff. He was all they needed.

Bernie was edging toward the big time on the trading end of his business. No longer working miles away from the action, he had moved down to 40 Exchange Place, just a block from the New York Stock Exchange. He was in range of his target: the doors of the castle. Tired of operating on the fringe, he and a group of other small outfits devised a strategy to unlock the NYSE's monopoly.

Computers were the key. The system of listing penny stocks on pink sheets made trading outside the big exchanges move in slow motion. Madoff wanted to get information out in real time and make it available to everybody. With computers starting to come of age, he pushed the envelope, tapping the Cincinnati Stock Exchange to host a "third market" and plunking down $250,000 to upgrade their computer system. On February 8, 1971, the new method of executing over-the-counter trades was launched. Madoff's group christened it the National Association of Securities Dealers Automated Quotations, or NASDAQ.

Now market makers were able to display their quotes on computer screens for a universe of traders to see. With a more efficient marketplace, people began to take over-the-counter stocks more seriously, and Madoff's business hit the big time. He was regaled as "the king of democratization" on Wall Street.

Years later, the people he grew up with would scratch their heads over how such an average intellect could attain such heights. It turned out that Bernie Madoff had remarkable gifts, ones that emerged only when he entered the world of finance. His public successes on Wall Street bestowed legitimacy on his operation and won him widespread respect. The kid who found his worth in business, selling sprinklers, once again won respect by making money.

But the world could see only his public accomplishments; Madoff's private investment business was ballooning virtually out of the sight of the people down on Wall Street who were applauding his wholesale trading operation. They knew nothing of his colleagues' investment sales activities on his behalf, or the pie-in-the-sky profit guarantees being peddled from Corona to the Catskills. It would be years before anyone thought to ask questions about these practices. No one was asking how Madoff or his minions could bankroll the returns they were guaranteeing. No one seemed to mind that there was scant paperwork to go with his orders, or question why receipts were being issued on scraps of paper. No one was putting these pieces together and questioning whether trades were being made at all.

The man who challenged the New York Stock Exchange was the new prince of Wall Street. He would later claim that it would be another twenty years before his business took a sharp turn to the dark side. But Madoff never took a turn. He sped recklessly down an icy road from his first day on the job.

3

The In Crowd

By the mid-1970s, Carl Shapiro had little to do and a lot of money to spend. Sensible dresses for the modern woman had made him spectacularly wealthy. Twenty-three years after founding Kay Windsor, the clothing mogul took his company public, and nine years later he sold it to the Vanity Fair Corporation for thousands of shares in stock—which promptly quadrupled in value. Bernie Madoff invested the fortune and sent his mentor's net worth into orbit.

The cotton king was considered Boston royalty now, and he seized the role of philanthropist with a zeal that put him in a league of his own. He showered so much cash on local institutions that students at Brandeis University could spend entire days studying only at centers and halls named after him.

Shapiro and his wife had to find new ways to spend their

time, and increasingly they did so in Palm Beach, the clubby Florida habitat of the millionaire class. They were part of a wave of prosperous Jews invading the blue-blood enclave en masse.

The island is an old movie come to life, a flawlessly beautiful resort town where the fabulously rich throw parties in their mansions and drive around in Bentleys to black-tie charity balls. It's a land of ladies who lunch, a throwback to an era when men made the money and women tended to charities. The Florida of retirees driving golf carts and tourists dressed in jogging suits is a universe away. Taste permeates the air. The women speak in whispers. The help treat them like royalty.

For years it had served as the exclusive province of American WASP society, ruled over by old money aristocrats like Marjorie Merriweather Post, founder of General Foods and America's richest woman; her daughter, Hollywood actress Dina Merrill; and fashion doyenne Lilly Pulitzer. The Gentiles greeted the new arrivals with a collective scowl. They locked the Jews out of the preeminent golf and country clubs and frowned on them at their lavish fundraising events. Galled by the emerging Jewish neighborhood south of Sloans Curve, they dubbed it the Gaza Strip. But Jewish multimillionaires like Shapiro stubbornly forged ahead despite the unabashed hostility, for they had the same aspiration that brought the WASPS there a century earlier: a desire to advertise their wealth.

The Palm Beach culture is a kind of homage to the European class system, marked by an obsession with money and an almost feverish desire for social superiority. It can be an absurd place, but it has proved impossibly alluring to wealthy people seeking social validation. For a generation of Jews raised by immigrants striving for a better life, their arrival in Palm Beach was the ultimate symbol of success.

Shunned by the anti-Semitic policies of the Everglades Club and the Bath & Tennis Club, Jews bought the Palm Beach Country Club and transformed it into their own clubhouse. Unwelcome at the annual Red Cross Ball, they congregated at the Discovery Ball for Boston's Dana-Farber Cancer Institute. And so it remained for decades: a religiously segregated little town united in its disdain for the middle class but divided by its disdain for one another.

By the time Bernie Madoff arrived in the mid-1980s, he was a major power at NASDAQ, soon to become its nonexecutive chairman, and a big man on Wall Street. He was accumulating the trappings of the New York upper class, summering at a sprawling oceanfront house in Montauk, Long Island, and soon to buy a 4,000-square-foot penthouse on Manhattan's Upper East Side. But he too must have realized that for a man who has reached a certain pinnacle of American business, there is truly nowhere better suited to be on display than Palm Beach.

His reputation preceded him, though not merely because of his rising profile on Wall Street. If all he had was a powerful job and a lot of money, he and Ruth would have been placed somewhere in the middle of the Palm Beach pecking order and forgotten. There were plenty of men on the island who were wealthier, more successful, and more exciting than Bernie Madoff. But Carl Shapiro had smoothed his entrance by regaling his fellow captains of industry with stories about the genius he'd discovered whose investing brilliance had made him millions of dollars. Virtually from the moment he walked in for a sandwich at the Palm Beach Grille, people looked upon him with awe.

With one of the planet's largest concentrations of the super-rich, the town is a natural hunting ground for investment advisors hungry for clients. But as a result of Shapiro's PR, Madoff didn't need to sell himself. The cult had already begun to form. At restaurants, his friends laughed at the stream of supplicants introducing themselves as he tried to eat. Some were subtle enough to simply shake his hand and refrain from asking him point blank to invest their money. Others would pop the question immediately. To the latter, Madoff would offer an impatient signal of disinterest. "Not now," he'd say, or "I'll think about it." His indifference to new business only added to his mystique.

There was an air of discretion about Madoff that embellished his image as an elusive guru. He never talked about money in public, never boasted about his success, never even talked on his cell phone in front of other people. When his phone rang on the golf course he'd walk away to answer it, telling his friends to go on to the next hole. "I never heard Bernie quote how much money somebody had," says a close friend. "He would never discuss if a client had a hundred thousand dollars or a hundred million dollars."

Yet there was something still uncultivated about Bernie Madoff after all these years. He had a Queens cabdriver's voice—he often *tawked like dis*—that was curiously out of synch with the dark and mysterious image he so cherished. He also had a terrible blinking problem that jibed badly with the mystique. Perhaps to compensate, he developed a deep interest in his appearance. He adopted a business tycoon's power look, his hair combed back from his receding hairline into a long, impeccably groomed gray mane. His weekend rituals included a Saturday afternoon stop at Trillion, a Worth Avenue men's boutique where the rich can pick up a $2,400 sweater between their golf game and an afternoon snack. Madoff purchased countless suits at up to $10,000 a pop when he wasn't shopping at Kilgour in London's Savile Row. "You could show him navy blue forever," said David Neff, the store's owner. He would linger at Trillion, where jazz played in the background, as he flipped through worsted spun cashmere suits, famous for their buttery-soft feel and smooth finish. He'd pick up a handful of Italian super 180 wool sweaters, fabulously expensive because of their unusual weaves. Like the suits he preferred, the sweaters were dark and conservative. "They didn't scream 'Bijon,' " Neff said. "They were just the best."

The town suited Bernie and Ruth in many ways. The island is separated from the rest of the world but for three mechanical bridges. There are no signs to Palm Beach on the highways, and to drive through its residential streets is to feel like a trespasser. Most of its mansions and exclusive condominiums sit behind dense, expensively manicured hedges, giving the place the feel of the exclusive club that it is. All that's missing is a "Keep Out" sign.

The exclusivity appealed to the Madoffs. They were a private, reclusive couple who for reasons unknown at the time chose to plant themselves at the center of the socialite galaxy. The charity circuit held no interest for them. They dined together several times a week at a casual Italian restaurant on Royal Poinciana Way named Cucina Dell'Arte. Though the restaurant featured a disco ball on the ceiling and ran ads inviting their clientele, "Talk Loud. Eat Well. Laugh Often," the two ate quietly off to the side, speaking to one another so discretely they were almost whispering. The number of close friends they had on the island could be counted on one hand, and the group kept to themselves, going to movies and occasionally renting a cabana together at The Breakers, the iconic Palm Beach hotel.

Friends loved the Madoffs' unaffected ways. Ruth was small, slender, and chic, with an effortless sense of style that contrasted with the pretentiousness of her neighbors. "You'd think she was a doll," says a friend. "She was the cutest thing, even with nothing special on. And if you went to the Palm Beach Country Club with them, or Club Colette, people would be decked out, practically wearing tiaras on their heads. Ruth, who could have anything she wanted, looked beautiful and elegant and tasteful." Bernie's secretary, Eleanor Squillari, said that Ruth "wanted to be perfect for him. She would never allow herself to gain weight or have a hair out of place, and she always kept an eagle eye on him, especially when he was around young, attractive women."

The couple had been inseparable since Far Rockaway High, and only grew closer over time. Bernie hated golf and was perpetually turning down invitations to play; they were thinly veiled attempts to get him to invest their money, he thought. But because Ruth loved the game they would play alone or occasionally with a small group of close friends. Even their quarrels reflected the couple's tight, codependent relationship. "If Bernie said something to Ruth that annoyed her, she'd say, 'Go fuck yourself,' or 'I don't give a shit,'" said Squillari. "That's the way they talked to each other."

Ruth was by far the more adventurous one. She loved to eat out, loved to shop, loved to do anything fun she read about in the papers. "I heard about this great restaurant, Bernie," she'd tell him. "Let's try it." Over dinner with friends, she was gregarious, chatty, and curious about what other people were eating. She'd eye their plates with her fork hovering, or sheepishly ask, "Can I have your bone?"

Bernie was a quiet and distant man, never the sparkplug at a party or a dinner table. He didn't like meeting new people and could grow aloof around friends. He was happiest when he was alone, puttering on his 55-foot 1969 Rybovich yacht, christened *Bull*. He had staff to take care of his boat, but he insisted on cleaning it himself, getting on his knees to scrub fish blood off the floor or washing mold off the boat's canvas coverings with some soap and a sponge. He was consumed with cleanliness.

He and Ruth purchased a beautiful home on the Intracoastal Waterway that was a perfect reflection of their uneasy relationship with Palm Beach; it sat among a row of spectacular mansions, yet was almost completely shrouded from view, as if purposely designed to be overlooked. For Bernie, the house was both a joy and an obsession. He was exact about the look of the place and easily angered at any sign of disorder. If the trees in their front yard weren't trimmed perfectly he became enraged. "You go on ahead," he'd tell his friends, frustration creeping into his voice, sending them on to lunch so he could wait for the gardener to show up and get a piece of his mind. He was forever straightening rugs, drapes, objects on his tables. "Bernie, get off it already," his friends would complain.

Though gruff and impatient, he wasn't the kind of rich man who would demean a waiter or abuse the help. When he invited his friends onto the *Bull*, he would personally serve them food and cocktails. When they boarded his private plane, he would get annoyed when his stewardess made a fuss over him.

He seemed like someone endlessly in search of peace and quiet. He would often grow cranky and retreat to a corner in a room. "Stop yapping!" he'd yell when Ruth was on the phone or gossiping with a friend. "What are you talking about, shoes?"

Friends sometimes wondered about his desire for solitude.

"Bernie always had a sense of worry about him," said one. When he seemed stressed, he would neither confide his problems nor explode over them; he would simply wander off, close his eyes, and take a nap. "He would fall asleep so fast," a friend said.

But Madoff's aversion to people didn't stand in the way of his mission. He had come to Palm Beach for a reason.

The drive to Fort Lauderdale from Palm Beach takes about an hour on I-95, and although it parallels the ocean from beginning to end, the experience feels a little like leaving Oz for the real world. One of three people on the streets of Fort Lauderdale is African American, a jolting sight after spending time in an all-white universe. Most of the people here make a lot less money and live in houses worth a whole lot less. But there are fabulously rich people in Fort Lauderdale too, like the couple in Bay Colony whose vast estate included a 6,000-square-foot house, a 10,000-square-foot pavilion for parties, an indoor swimming pool, and a cold storage compartment for fur coats. The owner of this particular house was making a fine living finding investment clients for Bernie Madoff.

Back in the early 1960s, Michael Bienes was a serious, bespectacled young accountant paid to crunch numbers in Saul Alpern's cramped office. Over the years, the entire company had been transformed into a cog in Madoff's vast investment machine, and Bienes and his partner, Frank Avellino, became spectacularly wealthy on the commissions they were earning for Madoff. In 1980, Bienes tossed his green eyeshade to the wind and took off for Florida. Casting his former life and CPA's image overboard (along with his Jewish identity, having converted to Catholicism), he capped his teeth, changed his hairstyle, and reinvented himself in the Sunshine State as the antithesis of a boring accountant: he was now a bon vivant, the best party host in Fort Lauderdale.

"How can you not feel good at a Bienes party?" a writer wondered in the Fort Lauderdale *Sun Sentinel*. "The food is always superb. The service, impeccable. You never worry where to dump that canapé toothpick or that empty glass. Someone always appears to whisk it away. Tuesday night, chefs filled your plate at the buffet, and an attendant carried it to your table for you. All you had to do was chew."

Bienes had morphed into Fort Lauderdale's answer to Jay Gatsby, lavishing excess on his friends at parties they would talk about for years. Lest anyone miss the point, he threw a "Gatsby Gala" in 1995, which sent the local society writers into raptures. "The pool was covered with a dance floor, Jerry Wayne played the music of long ago," wrote the *Sentinel*'s Martha Gross. "The hooch flowed like water, and flappers swarmed about. . . . The dinner was exquisite. Gold service plates and table wear, gold-tipped napkins, gold runners between the bowls of roses. You've got the picture—24 carats all the way."

When not doling out millions to charities, Bienes and his wife, Dianne, a former accountant herself, were decorating their $7 million estate with art worthy of a small museum.

They preened in front of their possessions for the *Miami Herald*, facing one another over "an Asprey Globe made of rock crystal, gold and diamonds." Years later, they flew to England to host a fundraising dinner at Windsor Castle with the Prince of Wales. "They're a flamboyant pair," reported a London arts publication. "They dress to kill, they like to party and they're legendarily good hosts." Bienes told the writer that he and his wife were "just two people who give money to things we like . . . by instinct, with emotion, preferably with some fun attached."

The fun was fueled by an astonishing sales operation for Bernie Madoff. An effort that began in the 1960s with some old folks at a creaky Catskills hotel had mushroomed into something far larger. By 1992, Avellino and Bienes had raised nearly half a billion dollars.

Their pitch hadn't changed from the days of the Sunny Oaks Hotel and Cottages. Investors were guaranteed that they'd realize an agreed-upon return on their money, often as high as 20 percent. It was hard to find another deal as good as that, and clients rushed in to board the money train. It was a happy experience for both seller and buyer. Investors were making the double-digit returns Avellino and Bienes had promised, and Madoff and his two star salesmen were getting rich in the process. It was "easy-peasy, like a money machine," Bienes said.

It took almost thirty years for the feds to catch wind of it.

In 1992, a skeptical investor contacted the Securities and Exchange Commission about Avellino and Bienes's spectacular returns. Government lawyers went to the files but couldn't find any documentation for the company. It wasn't registered to trade securities, and the two men held none of the required licenses. There were no regular filings reporting their trading activity, no annual reports issued to investors. For a while, Avellino and Bienes weren't even telling their clients how their money was being invested. When one complained, they told him he could take his money back.

Like Bernie Madoff's parents and later his father-in-law, Bernie's star salesmen were running an illegal securities operation. The difference was the gargantuan size of Avellino and Bienes's venture: the two accountants had collected checks for $440 million from 3,200 people. It was one of the largest sales of illegal securities in American history. Worse, the duo's guarantee of high returns had the smell of a scam. Word raced through the SEC that its staff had uncovered a massive Ponzi scheme, in which a con artist posing as an investment manager pays investors their returns from the cash of his newer clients and pockets the rest. Usually, such schemes collapse when the scam artist can't raise any more money. In the case of Avellino and Bienes, though, new money never stopped flowing in.

Investigators thought they were staring at a catastrophe and filed suit against the two in November 1992. To their surprise, though, Avellino and Bienes had the best excuse money could buy: Bernie Madoff had the cash.

In the span of a few days, Madoff refunded $440 million to Avellino and Bienes's investors. Apparently it never occurred to the SEC to ask why an investment manager would have so much cash on hand. Investigators never thought to investigate whether it was Madoff who was running a Ponzi scheme. After all, this was Bernie Madoff, a brand name on Wall Street and a founder of NASDAQ.

Clues to a hoax were everywhere. Avellino and Bienes had been guaranteeing stratospheric returns on people's money, even in years when the market was plunging. How could they make such a promise, especially if someone else was doing the investing? Avellino begged the question himself in his deposition: "If I was short and there was a shortfall," he said, "I would be in trouble." No sane man would take such a risk, unless he knew it really wasn't a risk at all.

"Who was the broker with the Midas touch?" asked the *Wall Street Journal* at the time. "The mystery broker turns out to be none other than Bernard L. Madoff—a highly successful and controversial figure on Wall Street, but until now not known as an ace money manager." When reporter Randall Smith of the *Wall Street Journal* reached Madoff on the phone, he was happy to talk. Madoff shrugged off the guarantees that the accountants had made to their clients. "I would be surprised if anybody thought that matching the S&P over 10 years was anything outstanding," he said.

But the real record of the stock market over those years was a very different story. From the end of 1960, when Madoff founded his firm, to late 1992, there were ten years in which the market went down, not up. The market average over the thirty-two years was 13 percent, but that was only an average. Unless Madoff had discovered a historic formula that made it impossible to lose on Wall Street, it would be hard to understand how anyone could pay annual returns of 10 percent or more when the market dropped at least 10 percent six times from 1962 to 1977.

Yet Madoff claimed that he had found just that secret. He gave the reporter a lesson in what he called his split-strike conversion strategy, an algorithm of stock and options trading that, he said, allowed him to limit losses and produce consistent returns whether the market moved up or down. He claimed that he was able to limit his risk by using options—which let him buy and sell shares at pre-set prices—and knowing when to go in and out of the market at just the right moment. His language was dense with intimidating financial-speak. "The basic strategy was to be long a broad-based portfolio of S&P securities and hedged with derivatives," he explained. As a result, he said, he could go for long stretches of time without losing money. It was an explanation that seduced his fans and disarmed his skeptics for decades. As for Avellino and Bienes raising money illegally, Madoff feigned ignorance, saying he thought they were registered all along.

Years later, Bienes would tell a different story. He said Madoff assured him that neither he nor Avellino needed to register with the SEC. "We had doubts, and we passed them on to Bernie in meetings," Bienes said. "And he said, 'Listen to me, okay? I know the biggest lawyers on Wall Street, and I've told them about this, and they say it's okay. You're just guys who work for my father-in-law. You're a client of my firm. That's all you are.' "

Asked why he didn't register to remain on the safe side, Bienes said it was out of the question. "You just can't do that. . . . Bernie didn't want us to. . . . We were always captive to him. He owned us."

After the SEC put a stop to Avellino and Bienes and Madoff returned all the money to investors, the inquiry lost steam. The two accountants made a deal to shut down their company, pay a $350,000 fine, and submit to an audit. That audit would have likely probed for evidence of Madoff's trades, but the two accountants played their overseers for fools. Avellino claimed to Price Waterhouse, the court-appointed auditor, that he'd never kept any books. "My experience has taught me to not commit any figures to scrutiny when, as in this case, it can be construed as 'bible' and subject to criticism," he said. Then, amazingly, he and his lawyer, Ira Sorkin, managed to basically kill the audit altogether after complaining to a judge about the money it was costing to comply with it. "I am not a cash cow and I will not be milked," Avellino said.

And with that, the federal government went away, leaving behind a river of leads and questions that hadn't been asked. Where was the evidence of Bernie Madoff's trades? Who was selling him the options fueling his unorthodox trading strategy? How could his key salespeople guarantee investment returns? Where were his books? The answers would have been easy to find.

But Madoff escaped the investigation unscathed, with his

firm and his reputation intact. Avellino and Bienes took the fall, and were lucky to get off so easily.

It wasn't over for Bienes, though. The end of his employment with Madoff was personal. He said he arranged a meeting with Avellino and Madoff on the nineteenth floor of Madoff's headquarters. Nancy Avellino and Dianne Bienes joined their husbands, hoping to repair the relationships that had so suddenly collapsed. The five of them sat inside a glass conference room, where civility was held together initially by a fragile peace. It was Bernie who killed the mood. "It's over now," he said. "You fucking guys have to get yourselves in order now."

The meeting called to end the storm quickly renewed its fierce and destructive intensity. The insults escalated into an assault, which Bienes intended to fight. He would not be demeaned by Madoff, especially in the presence of his wife. "You son of a bitch, it's over now!" he shouted. "We went through it. It cost us a lot of money and a lot of grief. And it's all your fault, Bernie. God damn you, it's your fault, because we asked you, 'Should we be registered? Should we get registered?' We were willing to do it. We were willing to pay any lawyer any fee. . . . And now you're looking at us as if we did something wrong?

"We came to you more than once and said, 'Are we okay? Are we doing something—?' And you assured us, big shot, that we were fine, we were just investors, when you knew God damn well we weren't."

Bienes had lost control. When he found a moment to

breathe, he noticed that Madoff was the only one in the room who had not so much as blinked during his tirade. He was sitting in his chair, arms folded. He finally spoke. "Look," he said. "I heard enough from you. Now I want you to stop. You're starting to get to me."

The words scared Bienes. No longer allowed to receive a cut of the business, he and Avellino were now just ordinary Madoff investors who could be cast off at any time. Madoff's grip on him was still solid. Bienes paused a moment. "Bernie, I'm sorry," he said. "I'm just a very scared person. Let's forget what I said and go on with this. I apologize."

In November 1992, shortly after the SEC shut down Avellino and Bienes's operation, Stephen Richards got a phone call from Bernie Madoff. The Ohio State frat boy who had met Bernie on Ruth's porch in Laurelton thirty-nine years earlier had invested money with Madoff through Avellino and Bienes and was alarmed at reports of their troubles with the SEC. By the time the phone rang, though, he had received a refund check for every dime of the millions of dollars in his account, just as Madoff had promised to the SEC.

"Stephen, it's Bernie Madoff. I just wanted to tell you that you don't have to worry." The big man wasn't just calling to reassure Richards. Madoff wanted to know if he was interested in sending the money *back* to him. Madoff said he would reinvest it and handle the account himself from then on. It was Richards's choice. "What do you want to do?"

Richards had retired in part because of the terrific returns he was making on his investments with Madoff. Despite the whiff of scandal hovering over him, Richards didn't want the train to slow. "Well, you know we expected you to keep doing this," Richards said.

"Just send me back the money and I'll open up new accounts for you," Madoff said. His tone was reassuring. Everything would be okay.

Richards did exactly that: he sent Madoff a check and reopened his account with him as if nothing had happened. Bernie was back in business.

With the doors to Avellino and Bienes's operation padlocked, Madoff needed another key money raiser for his investment advisory business. Palm Beach was filled with potential clients clamoring to get into his fund. It was important to Madoff that he seem disinterested, but someone had to actually raise money from these people.

He didn't have to look far for his man. Carl Shapiro offered up his son-in-law.

Robert Jaffe was a creature of high society, best and perhaps only understood by the members of his own plasticized culture. There he was, motoring down Worth Avenue in his green 1954 MG TF British convertible, virtually swallowed up inside its deep carriage. His hair was slicked back in the style of a European playboy, his suits were tailored, and his posture was straight as a ballet dancer's. His cultivation showed with every arch of his perfectly shaped eyebrows. He had always savored high style, working his way through college at the tony clothing store Louis Boston, where he helped businessmen with middle-aged paunches wriggle into their Brioni suits. Years later, when Jaffe's fortunes had risen, the company would name him its best customer. "The clothing I wear is more—dare I say—cutting edge," he told the men's fashion magazine *DNR*, which profiled him in 2008. "It's a few years ahead of the pack. Once you've had filet mignon, you don't want to go back."

Jaffe made his fortune the Palm Beach way: he married into it. Ellen Shapiro, Carl's daughter, gave Jaffe entrée into Palm Beach society and the life he was destined to lead, that of a rich dandy. He sat on the boards of Palm Beach charity balls, ate at the best restaurants, perfected his golf game, and wrote fat checks to the causes of his wealthy friends.

His entrance into the Shapiro family's world was by extension an entrance into Bernie Madoff's as well. They were an odd duo, this peacock of a man, in the words of social observer Lawrence Leamer, and the street-smart Queens kid with an outer borough accent. But Madoff and Jaffe made beautiful music together, for they perfectly served each other's needs. In Jaffe, Madoff found a socialite who was out seven nights a week, the ideal man to spread the word among the jet set of Madoff's success. And Jaffe found a golden source of power: access to Bernie Madoff.

Everyone wanted in on Madoff's fund, so revered was his name. Jaffe became the man whose ring had to be kissed to get there. He was Carl Shapiro's son-in-law, but when he purchased his spectacular $7.8 million waterfront property, it wasn't near the old man. It was two doors down from Bernie on North Lake Way.

Jaffe became a vice president of a Madoff company offshoot, Cohmad Securities Corp. (the "mad" in the name standing for "Madoff") and spent his days connecting eager investors with the all-powerful investment guru. Jaffe's friends assumed he was doing them a favor by sending them to Madoff. In reality, Bernie was paying Jaffe around 2 percent of each deal.

The most fertile hunting ground for new customers was the Palm Beach Country Club, a yellow, plantation-style estate surrounded by palm trees, a lush golf course, and car parkers in white shirts and shorts running through a parking lot filled with Porsches, Mercedes, Bentleys, and Jaguars. The Club's $300,000 initiation fee was just one hurdle to admission. Applicants had to prove their good character and standing in the community, and their charitable contributions to Jewish organizations had to be in the six figures. Once admitted, members had the privilege of wandering through a facility resembling a fusty old university club, replete with traditional silver chandeliers, trophy cases with plates inscribed with names of winning bridge tournament winners, and old men wearing bright red jackets.

Bernie would come for an occasional lunch at the Club, wearing a Florida tan and dressed in chinos and a blue blazer. Invariably he'd eat alone with Ruth, her hair colored a soft blonde and her outfits à la Nancy Reagan, fashionable but not flashy. The two were just another prosperous Palm Beach couple, yet members at the tables they passed by tended to stare and whisper.

It was Jaffe who did the hard work of schmoozing Madoff's clients. He was once the Club's champion golfer, a tip-off to the time he was spending there. As the years passed, perhaps a third of the Club's members would invest money with Madoff, maybe more. They included Jerome Fisher, owner of the Nine West luxury shoe empire, and Irwin Levy, founder of South Florida's famous Century Village retirement homes. Some of the Club's members were deeply involved in Jewish charities. For years, Robert Lappin, a developer of the Salem, Massachusetts waterfront, offered trips to Israel for every Jewish child in Salem. He placed the foundation's endowment in Madoff's hands.

Unlike the senior citizens recruited by Madoff's father-in-law and the thousands who flocked to Avellino and Bienes, the wealthy of the Palm Beach Country Club were attracted less by Madoff's high returns—by now he was producing 10 to 15 percent yields, down from 20 percent—than by his consistency. Madoff had weathered up markets and down markets for decades, but his results rarely changed. That kind of performance was highly valued by moguls looking for safe places for their millions. At the Club, Madoff's name became synonymous with trust. He was known for investing only in Coca-Cola, Exxon, McDonald's, and other blue chip companies. And he was a member of the tribe. Jews of his generation were brought up to think of other Jews as extended family members, with a shared responsibility to look out for one another. They felt more comfortable going to Jewish doctors, Jewish lawyers, Jewish accountants. Madoff became known as "the Jewish T-bill." Their money was as safe with him as with the U.S. Treasury.

The values of the Club's members were borne of a thousand childhoods spent in middle-class Jewish neighborhoods like Laurelton, Queens. "The Jewish world was very tight and highly networked," recalled Irene Shapiro, a contemporary of Madoff's back in Laurelton. "The Kleins knew the Solomans, who knew the Goldenbergs, who knew the Steinbergs, who knew the Hoffners. Everything was done through 'the phone call.' My brother wanted to go into retailing, so a call was made to Mr. Temona on the board of Lerners. I wanted to transfer to NYU and live in the one dorm: a call was made to Professor Levine."

"There was another type of man in this world," she said. "He was the man who knew a lot of the others and could get you into contact. He was really [the] one on top. To be the man who was called was a big deal." Shapiro believes that Madoff "wanted to be one of them, the big shot, the Jewish prince who could dole out favors and advice and clear the path for others, the savior and a source of great power and great admiration. He wanted to be a power in the Jewish world, the world of his fathers."

Some of the most powerful men and women in corporate America were competing to get close to Bernie. Among the rich and powerful, he had an aura of royalty. He may have been known as the Jewish T-bill instead of the Jewish prince, but it was the triumph of a lifetime.

Madoff could not have found a more lucrative source of cash for his investment advisory empire than this elite and insular community. But the benefits seemed psychic as well. The Country Club was to Palm Beach what the Laurelton Jewish Center was to Madoff's boyhood neighborhood, a beacon of wealth and pride for Jewish residents like him. Madoff, who grew up feeling inferior, may well have looked on both institutions in the same light: as societies of judgmental Jews he wanted to impress—or get his revenge on.

Sometime in the early 1990s, a Palm Beach accountant named Richard Rampell picked up the phone and placed a call to Bernie Madoff in New York. Sitting at his desk inside a 17,000-square-foot building he owned, Rampell had some questions. The two men had never met, but Rampell owned Palm Beach's largest accounting firm, and a wealthy client of his had come to him with an unusual request.

A member of the Palm Beach Country Club, the client had invested $5 million with Madoff on Carl Shapiro's advice back in the 1980s. Five years later his money had doubled. The stock market had plunged, but his Madoff account was still earning double-digit returns. "How's this guy making money when no one else is?" Rampell's client asked him suspiciously.

Rampell, professorial-looking in his signature bowtie, studied his client's statements from Madoff. The first thing he noticed was how consistent the returns were: no matter what the market conditions, the returns were always in the 15 percent range. The second was that Madoff was showing hundreds of trades in every statement, highly unusual for a broker.

Madoff took his call. "My client is concerned about what exactly you're doing here," Rampell said. "Is everything okay?"

Madoff was polite but to the point. "I don't reveal my trading strategy," he said. "It's proprietary. We have our own trading program."

"But I'll tell you this," he added, describing his split-strike strategy, "I can make money when the market goes up and I can make money when the market goes down. I cannot make money when the market is flat."

Rampell ended the conversation convinced that Madoff was hiding something. His strategy required an impossible level of precision to succeed. And his winning percentage was far too consistent. No broker on Wall Street performed that flawlessly.

He played back his doubts to his wealthy client. "I've never seen any portfolio manager who's never had a down year," he told him. Madoff was engaged in insider trading, he speculated, using confidential information gleaned from the market-trading half of his business (front-running, in Wall Street-speak).

His client thought about it for a moment. He had come to Rampell because he was dubious of Madoff's methods. But he was pulling out a million dollars a year from the enormous returns Madoff was producing for him, and the implications of shutting off the spigot began to sink in. Rampell's analysis suddenly seemed like an unwelcome intrusion. "I don't want to hear anything about it," the client said. And with that, he returned to the life of luxury Bernie Madoff had given him.

4

Friends and Enemies

Every financial house has a back office. You can sense you're in one by the quiet in the air; there are no star investment bankers racing self-importantly through corridors, no high-strung traders shouting orders, no crowds of stressed-out staffers circling the copy machines. Back office workers sit in quiet cubicles and small offices implementing other people's decisions. Not all are paper-pushers; some are mathematicians with sharp, sophisticated quantitative skills paid to design impossibly complicated computer programs carrying out the company's strategies. On Wall Street, they're known as rocket scientists.

Harry Markopolos was a classic rocket scientist: unmistakably brilliant, with a talent for understanding the algorithms of stock trading and conceiving trading models designed to beat the market. True to the stereotype of a math genius, he was pale and bookish, with a thinning comb-over and big, round glasses. He was a serious man, with a smoldering intensity that made him appear a little impatient and even a bit angry.

He had the air of a spy about him. He was a graduate of the Army's Special Warfare Center at Fort Bragg, where he'd swum fifty meters in combat boots and battle uniform, and then commanded a special operations squad in Europe and Africa for seven years. He was a reserve officer, working in civil affairs—the Army's "soft power," as he put it—organizing public health clinics instead of kicking down doors and jumping out of helicopters. But the allure of intelligence seemed to capture his imagination; he used terms like *field operatives* and *intelligence networks* in his business conversations, a little over the top for a guy writing trading programs. It was one of the reasons he struck colleagues as a little eccentric.

But he seemed born to the task he assumed in 1999.

His employer, Rampart Investment Management in Boston, was a fairly small shop specializing in options trading, which happened to be the strategy of choice for Bernie Madoff. The two companies often competed for clients. Markopolos had once designed a split-strike conversion program for Rampart that was similar to Madoff's, but it hadn't been greeted by euphoric fans as Wall Street's holy grail, and he hadn't been anointed a mysterious but revered investment guru. Markopolos's product sputtered along, producing small, unspectacular returns, until his clients lost interest in it and the company dumped the product.

Not surprisingly, it galled him when the marketing guys at Rampart threw Madoff's success in his face, asking him in smart-alecky tones why he couldn't match his returns. The irritant became something a lot more when Rampart's marketing chief, Frank Casey, traveled to the twenty-second floor of a Madison Avenue office building in Midtown Manhattan in the winter of 1999. He came to pay a sales call on a friend of Markopolos's who worked at Access International Advisors, an investment company catering to European royalty. The outfit had a portfolio reaching into the billions.

Casey, a former Army infantry captain himself and a bull of a man, was led into the office of Access's president, a dapper French aristocrat named René-Thierry Magon de la Villehuchet. He was a regal presence, 65 years old and the picture of French nobility. His Madison Avenue office was filled with fellow members of the European jet set, such as Philippe Junot, the former husband of Princess Caroline of Monaco, and Crown Prince Michael of Yugoslavia. There, they solicited the investments of European dukes and princes who entrusted them to keep their money safe and their returns high. Increasingly, they were passing the cash along to Bernie Madoff to manage for them.

The two silver-haired men sat down, one a plain-spoken former military man, the other an elegantly dressed French nobleman who spent his spare time living in his family's castle in Brittany. De la Villehuchet listened politely to Casey's pitch on behalf of Rampart, one of the country's leading asset managers specializing in option strategies. But it soon became apparent that the Frenchman was in love with his Madoff account. De la Villehuchet had met Madoff in the mid-1980s through his business partner, Patrick Littaye, and was impressed with the American's wisdom and common sense. The longer he stayed with Madoff, the more impressed he became; Madoff was spinning a profit of 12 to 15 percent, with few losing months. It was a phenomenally successful record, far better than anything in Casey's briefcase.

Chagrinned, Casey asked de la Villehuchet how well he'd vetted Madoff. "I've been doing due diligence on this fellow," he answered. "I get reports every day of which positions are bought, which are sold, and which options are purchased and which are sold." De la Villehuchet was accepting Madoff's information without question and having three of his clerks diligently enter it into the Access International computer system.

Casey was amazed. Investment managers are required to keep their assets with third parties so auditors can ascertain that they're real. But Madoff wasn't registered as an investment advisor, and was holding the money himself. "Why are you bothering logging in all these pieces of paper?" Casey asked de la Villehuchet. "He can make that up. There's no check and balance here."

But de la Villehuchet had grown increasingly comfortable with Madoff as time went on. His operation was becoming a full-blown feeder fund for Madoff, a go-between for wealthy European investors and the craggy Queens native who was working his magic for them. Casey was frustrated with the apparent superiority of Madoff's returns, but also unsettled that such a big investor wasn't being more skeptical of his methods.

De la Villehuchet seemed almost too refined to spend his time dealing with money, and in fact had eschewed business school as a young man in favor of a PhD in economics from Sciences Po in Paris. He was a gentleman banker living in an era in which there was no such thing. He came from a world in which honor was paramount and a man's word was close to sacred. His brother, Bertrand, felt he was a little naïve. "Once my brother trusted someone, he never went back on it," he'd later say.

Unmoved by Casey's arguments, De la Villehuchet shook his hand and politely ushered him out of his office, unaware that his trust in Madoff would cost him his life.

When Casey arrived back at Rampart Investment in Boston, he headed straight for Harry Markopolos's office and plunked down a copy of de la Villehuchet's returns. "Harry, why can't you do this? You're an options guru," he said. It was a good-natured challenge, but the question lingered in the air. Markopolos's competitor was wiping the floor with him.

Casey subsequently got hold of a marketing flyer from Broyhill Securities' All-Weather Fund. The mutual fund detailed seven years of monthly returns, and Casey had learned that the unnamed manager of the fund was Bernie Madoff. Markopolos studied the results. Almost every month showed a profit. The graph of the returns formed a 45-degree line heading straight up—for seven years.

Markopolos's own split-strike formula beat the market some months and lost to it in others. That was natural. No one could beat it all the time.

"This has gotta be a fraud, Frank," Markopolos said. "I mean, look at the straight line return. You know options, you know there's a relationship to the world marketplace. There has to be a correlation to down markets. There's none here." He asked for some time to study Madoff's returns further. He was on a mission.

He put all of Madoff's numbers into a spreadsheet, then compared them with the Standard & Poor's 100 Index, the blue-chip stocks Madoff was purporting to be buying and selling for his clients, together with the options that protected him from steep losses. The first thing Markopolos found was that some of the stocks in Madoff's statements weren't even in the S&P 100. They were in the larger S&P 500. It was a red flag.

He compared Madoff's performance buying and selling those stocks against their performance on the open market. Out of eighty-seven months, the S&P had been down twenty-eight of them. Madoff's statements showed that he was down just three. "No money manager is only down 3.4 percent of the time," Markopolos later concluded. "That would be equivalent to a major league baseball player batting .966."

He tried another test, opening the "Money and Investing" section of the *Wall Street Journal* and scanning for the number of options contracts being traded that day. There weren't enough options in existence for what Madoff was claiming to be doing.

Finally, to double-check those findings, Markopolos and a colleague got on the phone and called around to Chicago options traders. None was doing business with Madoff.

It reeked of a scam.

Markopolos was surprised enough that money managers like de la Villehuchet were investing 100 percent of their clients' money in any one fund. To Markopolos, betting it all on Madoff was crazy. "Bernie Madoff was a 'no-brainer,'" Markopolos recalled, "but only in the sense that you had to have no brains whatsoever to invest into such an unbelievable performance record that bears no resemblance to any other investment manager's track record throughout recorded human history."

Checking his instincts, he walked across the street to the fifth-floor offices of Northfield Information Services and paid a visit to its president, Dan diBartolomeo. Markopolos had been a student of his in a class on quantitative methods for investment analysis, and he considered diBartolomeo one of the world's great financial mathematicians. He was a mentor to other rocket scientists, a physicist who applied high-concept principals to the market, comparing the fluctuation of stock and option prices in one study to the physics of heat and how it spreads. It was genius-level stuff, and companies paid him a handsome fee to learn the science of making money and limiting risk in the market.

Markopolos presented him with the information he'd been studying and left him alone with it for a day. DiBartolomeo then went about trying to reproduce Madoff's results. He designed a split-strike model and plugged in three years of Madoff's trades. He failed each time to come up with the returns Madoff was claiming.

He called Markopolos. "The numbers are too good to be true," he said.

They concluded that Madoff was either front-running or operating an elaborate Ponzi scheme. Front-running is a form of insider trading; Madoff could have been using information about pending trades from his firm's trading desk to buy or sell stock before the market reacted to those trades. A Ponzi scheme is almost as simple. Money from new clients is used to pay off older clients, and the swindler takes money for himself off the top. Everyone is happy until the schemer runs out of new money and can't pay off his older clients any more. Madoff was regularly honoring tens of millions of dollars in redemptions, or withdrawals, at the time; if he were running a Ponzi scheme he'd need massive new cash infusions to replenish it. As it happened, money was flowing into his offices as fast as he could count it.

The implications for investors and the financial world were breathtaking. If Madoff was exposed as a crook, it would constitute one of the biggest financial frauds in history. Yet the implications for Harry Markopolos were also huge. The government pays whistle-blowers a lot of money for exposing financial crimes. Federal bounties usually reach into the thousands, and sometimes hundreds of thousands of dollars. But the reward is determined by the size of the scam, and Madoff was trading more than 5 percent of the total volume on the New York Stock Exchange. If Markopolos were able to bring regulators the head of Bernie Madoff, it could make him a millionaire.

In the dead of winter, arctic blizzards pummel Minneapolis with wind chills reaching 30 below. Businessmen and women navigate downtown through enclosed bridges that make it possible to walk for miles without having to brave the elements. It's a city that prides itself for being prepared. Yet nothing prepared it for Bernie Madoff.

In the early years of the twenty-first century, Marquette Asset Management handled the wealth and investments of the most prosperous families in Minneapolis; no one passed through its doors without at least $5 million to invest. One morning, an enormously wealthy client sat down with the company's managing director and informed him that he was pulling millions out of his account.

Sitting behind his desk in his banker blue suit and Brooks Brothers tie, John Pohlad wore a look of resignation as he took in the bad news. He prided himself on his rectitude, his conservative image, and his family's good name. Pohlads had served Minneapolis residents for decades. They were a respected brand in the Twin Cities, owners of the Minnesota Twins and a handful of venerable banking companies.

Minnesotans don't admire showy people, particularly

showy rich people. Money is a private matter there, something to be made and saved. So when wealthy residents of Minneapolis started chattering about Bernie Madoff, it was as though a contagion had infected the city's midwestern values. Clients who wouldn't think to discuss their finances were throwing discretion out the window and bragging that they'd worked a connection and gotten in with the legendary New York investor. It became a status symbol.

That morning, the son of one of the best-known families in the Twin Cities delivered the news to Pohlad. His family had decided to create a foundation to aid the Jewish organizations of Minneapolis and was kicking it off with a huge seven-figure cash infusion. It would all be invested by Bernie Madoff. "I know someone who was able to get us in," Pohlad's client told him. His attorney had a friend who said he was willing to turn a favor and get the foundation into Madoff's fund. It was a conservative move, the attorney felt, for Madoff invested only in blue-chip stocks and his returns were stable. He was as safe as a bond. "I've got a number of friends in this," the client said. "They tell me this guy delivers ten percent—at least—in good markets or bad markets. He's not just a money manager who preserves money. He can grow it."

It was a familiar refrain. Pohlad's clients were people with high net worth and investments averaging over $10 million. But they were increasingly throwing their money into something they couldn't comprehend.

Pohlad had never met Bernie Madoff, but in 2001 he began learning about his New York–based competitor. Madoff had been cultivating a small collection of Minnesota clients, most of them Jewish, since the 1980s, and they'd increasingly converted their friends and families into Madoff followers.

When Pohlad and his partner looked through Madoff's statements, the first thing they noticed was the poor quality of his paperwork. Asset managers who serve millionaires spend small fortunes producing elaborate color reports and statements; Madoff's were black and white and spit out by an old-fashioned dot-matrix printer, with sheets of lines showing little but stock names and dollar figures. None of the CPAs and MBAs in Pohlad's office could make sense of them.

"Do me a favor," Pohlad asked his client that morning. "Explain to me what Madoff does."

"It's a split-strike thing," he said. "They use options. They do hedging."

"What does that mean?"

"I don't know," he said. "But it seems to work." He was wagering his family fortune on the reputation of a man he'd never met and whose methods he barely understood. There wasn't much more to say.

Pohlad's office was twelve hundred miles from Madoff Investment Securities, but Madoff was casting his spell over rich Jews in cities across America. In the Twin Cities, a stockbroker named Mike Engler played the Robert Jaffe role, chatting up Madoff's name on the golf course of the Oak Ridge Country Club (founded for those "who knew the difference between a golf ball and a matzo ball") and at charity balls for Jewish causes. The parallels to the Palm Beach Country Club were eerie: wealthy Jews who had once been frozen out of mainstream society had formed their own elite institution, and Madoff became the go-to investment guy for those lucky enough to be allowed in. The only difference was the weather.

It was hard to know what to call people like Engler and Jaffe. They weren't salesmen; many of the investors they lured thought they were just golfing buddies passing along a good tip. They were more like operatives. But whatever they were, they were multiplying like franchises across America, using a game plan that might as well have been written into a manual. They were usually prominent members of the Jewish community, working out of exclusive country clubs. While investors thought they were just friends passing along a good tip, they were actually getting a percentage of the business. And they always portrayed Madoff as a closed fund, sparking a clamor to get in.

Madoff had created a spectacular image for himself. When he visited the Palm Beach or Oak Ridge Country Club, he was received like visiting royalty, mysterious and unapproachable. Always he kept a discrete distance from the people who pursued him, dining alone with his wife or with a local operative like Jaffe or Engler.

The two operatives raised hundreds of millions of dollars for Madoff in the 1990s. But they were just two of his many alter egos. Another was pounding the pavement of perhaps the most important square mile of wealth and power in America: Manhattan's Upper East Side. It was Madoff's home turf, but no one scrounged up cash for him as well as a burly intellectual named Ezra Merkin.

There are synagogues in Manhattan that are as grand as cathedrals, with giant stained-glass windows to rival the most majestic churches. By comparison, the Fifth Avenue Synagogue, catering to perhaps three hundred congregants on a Friday night, is the kind of building you can miss walking down the street if you turn your head. Percival Goodman, its architect, designed a solemn exterior with six vertical rows of oblong holes that are often compared to cats' eyes, though they look more like raindrops, or tears.

It isn't the building that makes the synagogue a singular institution in New York, but the people who fill it on Shabbat. Revlon chairman Ron Perelman, one of New York's wealthiest men and ex-husband of actress Ellen Barkin, sweeps inside on weekends escorted by a squad of plainclothes security guards. Synagogue chairman Ira Rennert, the junk bond king famed for building a twenty-nine-bedroom, thirty-nine-bathroom estate in the Hamptons, attends as well. Sitting in a place of honor on the bema near the rabbi is Elie Wiesel, the Nobel prize-winning author and Holocaust survivor.

Limousines and town cars would ordinarily stream down Fifth Avenue outside an event attended by such men of fortune and history. But the titans who attend services here are Orthodox Jews, forbidden by Jewish law from driving cars or even pushing an elevator button on the Sabbath, so they walk

from their homes. Some employ *shabbas goys*, maids or but-
lers whose job descriptions include pushing buttons, flicking
on light switches, and answering phones for them on Friday
nights and Saturdays.

The synagogue is a deeply traditional place, founded in
the late 1950s by congregants of another temple who revolted
when it committed the sacrilege of allowing women to sit
with men. To this day, women at the Fifth Avenue Synagogue
sit in a balcony above the men worshipping on the main floor.
Many members are staunch supporters of the Israeli right,
making it a friendly pit stop for Benjamin Netanyahu and a
handful of other Israeli politicians.

The synagogue's one-time president, Ezra Merkin, both
impressed and intimidated congregation members with his
deep reserve of religious training and faith in his own opinions. His considerable power emanated not just from his title
but from his bloodline, for his father, Hermann Merkin, was
a founder of the synagogue and a revered figure. Many congregants had known Ezra as a boy and had watched him
grow up to assume his father's mantle.

Ezra was round, bearded, and bespectacled, with a pro-
fessorial bearing burnished at Columbia and Harvard Law.
He was a product of Upper East Side Jewish society, raised
in the shadow of his father, a cold and remote German immigrant who'd made a fortune on Wall Street. As a boy,
Ezra was expected to rise at the dinner table once a week and
speak before his family about his Talmud readings. His sister,
Daphne, chronicled the event in a work of autobiographi-

cal fiction titled *Enchantment*, whose main character has a brother named Benjamin.

Benjamin slides his chair back on the carpet and stands up. He is good at his appointed task; he pulls on his lower lip and adjusts his glasses like a professor, ostentatiously studious. Every Friday night, Benjamin is called upon to trot out his erudition with an entertaining tidbit from the weekly portion of the Torah that will be read in *shul* the next day.

"There is a question," my brother says in his best pedantic singsong, "in this week's *parsha* about the distribution . . ."

Merkin's value to Madoff was priceless. Bernie and Ruth lived on the Upper East Side but were not *of* the Upper East Side. They didn't spend their evenings talking about politics or the arts at glittering dinner parties on Park or Fifth Avenue. Their vast penthouse apartment was on Lexington Avenue, maybe a hundred yards from the serious money crowd but a world away. The Madoffs could easily have afforded a Fifth Avenue spread, but they weren't that kind of couple; their decision to live closer to the proletariat on Lexington seemed more of a cultural statement.

Many of their closest friends were outer borough types, some of them former garmentos in Carl Shapiro's orbit who had made it big in the business world. Some lived in the suburbs, far from the elitist playgrounds favored by Manhattan's rich and overeducated. The Madoffs even maintained the modest house in Roslyn, Long Island, in which they had raised their family.

Merkin was an altogether different story. He was a proud member of the New York intelligentsia, with a formidable pedigree: president of the Fifth Avenue Synagogue, carrier of a famous name, educated in the Ivy League, former attorney at the prestigious law firm Milbank Tweed. When his fortunes swelled, he would plunge even deeper into the bosom of New York society by spending $11 million to buy a duplex at New York's storied 740 Park Avenue. Jackie Kennedy and John D. Rockefeller once called Merkin's building home; it was dubbed by author Michael Gross "the world's richest apartment building." Merkin would plaster his walls with his vast assortment of Rothkos, a private collection considered to be the world's largest.

Law held little interest for Merkin, and early on in his career he followed in his father's footsteps by entering the investment business. His was a curious niche, however. He formed a hedge fund, Gabriel Capital LP, for which he raised money but did no investing. That job was farmed out to others. Merkin, an investor would later say, was "a glorified mailbox."

He had a knack for picking the wrong business partners. The first investment manager he latched onto, Victor Teicher, made Merkin a wealthy man from his double-digit returns, until the party ended with his indictment on insider trading charges.

Merkin's next professional crush was on Bernie Madoff.

The two men found one another in the late 1980s, a connection that may have been greased years earlier by Merkin's father, an acquaintance of Madoff's. Their business relationship flowered in 1990, and two years later Merkin founded a full-blown feeder fund for his new associate. Madoff proceeded to produce double-digit returns for the fund, Ascot Partners LP, with marvelous consistency. By 2000 Merkin was investing a third of the assets from two more of his funds with Madoff.

As the years went on, the crush blossomed into a romance. Merkin gave Madoff entrée into his world, ushering him onto the boards of Yeshiva University and the American Jewish Congress. Luminaries of the Fifth Avenue Synagogue such as Rennert and *New York Daily News* publisher Mort Zuckerman wrote checks to Merkin's fund, unaware that he wasn't managing their investments at all. Ultimately, more than thirty Jewish charities poured their money into Merkin's feeder funds, a gusher of cash that eventually reached into the billions.

Madoff and Merkin's relationships with Jewish organizations grew so deep that it was often impossible to distinguish between their charitable work and their business dealings. In hindsight, there wasn't much that was charitable at all. Merkin rarely offered to manage money for free; his lifestyle was supported by the fees he charged for overseeing clients' investments. Whether it was a single investor or a nonprofit organization, Merkin charged 1 to 1.5 percent of the money he was asked to manage. Institutions such as the Ramaz School, a private Orthodox Jewish prep school on the East Side, paid Merkin—an alumnus—more than half a million dollars over fifteen years for his investment expertise. Further uptown, he served as chairman of Yeshiva University's investment committee, which turned over nearly $15 million in endowment money for him to manage. Over the years, he collected over $10 million in fees for his efforts. The sum is doubly curious since Merkin was simply passing the money along to Madoff, who also served on the Yeshiva board.

In the late 1980s, Merkin arranged for Madoff to sit down with one of the renowned figures of the twentieth century, the novelist Elie Wiesel. Perhaps the most famous writer to emerge from the Holocaust, he seemed to carry on his shoulders the sorrow of a horrified world in the wake of the disclosure of the Nazi death camps. He was an international treasure, winning the Presidential Medal of Freedom, the Congressional Gold Medal, and the Nobel Peace Prize.

"Look, you work so hard, what are you doing with your money?" Merkin asked Wiesel.

"I don't know," Wiesel said. "Shares here and there." He admittedly knew little of finance.

Merkin tantalized him with tales of the Jewish T-bill, a benevolent man who, though conservative with money, was able to generate piles of it. Madoff's fund was closed, and Wiesel wasn't rich enough for Madoff to care about, but Merkin boasted that he had an in with Madoff. A dinner was arranged so Wiesel could meet Madoff. The subject of money never came up. Instead, they talked about ethics and education, subjects dear to Wiesel's heart. Madoff made him an offer: he wanted to lure Wiesel away from Boston University to Queens College, Ruth's alma mater, where Madoff offered to endow a special chair in Wiesel's name. The author, flattered but loyal to BU, politely declined.

It was a performance that impressed Wiesel, who had already been enticed by the glowing things people said about Bernie Madoff. "He was like a God," Wiesel would say. "There was a myth that he created around him. That everything was so special, so unique. It was like a mystical mythology that nobody could understand. Just as the myth of exclusivity, he gave the impression that maybe a hundred people belonged to his club."

Wiesel wanted in. He and his wife, Marion, started out by investing their own money. Slowly, over time, it all went to Madoff: residuals from Wiesel's books, money from lecture fees, his Boston University salary—every nickel he'd saved over half a century.

And then they invested even more. Their foundation to help Jewish causes had raised $15 million. The money went to causes close to their hearts, such as a program to care for Ethiopian children in Israel. Madoff got it all.

Madoff's tear though the Jewish world was all the more remarkable because he wasn't a particularly observant Jew. He rarely went to synagogue, didn't engage in religious rituals, and rarely talked about his beliefs. In 1996, Ruth coauthored a cookbook titled *The Great Chefs of America Cook*

Kosher: Over 175 Recipes from America's Greatest Restaurants.
Inside was a photo of her standing in front of a dishwasher
wearing an apron; it gave some the impression that the Madoffs kept a kosher home. But Bernie loved pork sausages,
and Ruth had paid a ghostwriter to produce what was es-
sentially a vanity project, designed to give her something to
do. There was nothing kosher about the Madoff home or the
Madoffs.

Yet Bernie charmed people with his exaggerated stories
about his upbringing on the Lower East Side and his grandfather's bathhouse, and he had the demeanor of a wise Jewish
elder. Many referred to him as "Uncle Bernie." His Jewish
investors seemed to think of him as Jewish enough.

Some asked questions every now and then. The American
Jewish Congress had $21 million of its endowment invested
with Madoff, and one of its top executives heard scuttlebutt
from time to time that he was front-running his investments.
He asked the charity's controller to check out the reports. Yet
the probe went nowhere; it would have been impossible to
verify Madoff's veracity without sorting through hundreds,
if not thousands of his trading tickets, the controller concluded.

Madoff was benefitting wildly from his alliance with
Merkin. The synagogue president was the epitome of in-
tegrity. He was an intellectual who seemed somehow above
the nitty-gritty of dealing with money; he seemed more interested in raising it than investing it. His imprimatur was a
powerful lure for potential investors. "I'd put my own mother

in it," he told an investor about Ascot, the Madoff feeder fund.

Merkin eventually entrusted $2.4 billion of other people's money to Madoff. Yet Merkin wasn't putting much of his own money into the fund at all. Of the $470 million in fees he earned from Madoff, Merkin allegedly invested just $9 million back into the fund he was so successfully peddling.

It was a good time for investment managers like Madoff and Merkin. There was so much money coursing through the veins of the American economy in the late 1990s that they barely had to break a sweat in their search for new business.

In February 1994, Stanford University students David Filo and Jerry Yang launched Yahoo, after originally naming it "Jerry's Guide to the World Wide Web." Five years later, its stock was trading at $414 per share and Wall Street valued the company at $110 billion, more than twice the worth of General Motors. The dot-com boom spread talk of a "new economy" in America, in which stock prices no longer had to reflect a company's earnings. Tech companies like Yahoo and Amazon were exploding in value and vaulting to the center of the American cultural universe. Bill Gates and Steve Jobs were magazine cover boys and new corporate icons.

The hub of the mania was NASDAQ, the market Bernie Madoff helped create. The early dot-com startups needed a cheaper, smaller market, just what Madoff had needed as a rebel market maker forty years earlier. Now NASDAQ was doubling in volume and had a Studio 54-like cachet. And Madoff was the NASDAQ king. His market-making operation became the largest on the exchange by the new millennium.

Madoff's lower-key investment advisory business benefited as well. It was basically a hedge fund, but many on Wall Street had no idea it even existed. Those who did, though, trusted Madoff to handle the staggering profits they were making on their high-risk Internet investments.

Throughout the 1990s, the annual returns Madoff was spitting out dropped little by little. The investor who was once paid 20 percent returns was now receiving half of that. It wasn't clear at the time why this would be, but the more conservative returns served as a powerful lure to new investors. Bernie Madoff was safe. He was conservative. He was paying only 10 percent. They weren't go-go returns, but they were reliable, steady, a source of comfort. Like a T-bill.

In May 2000, five months after Frank Casey met with Thierry de la Villehuchet in Manhattan, Markopolos was given his moment to expose Bernie Madoff as a crook. He'd made an ally of Ed Manion, a staffer at the Boston office of the Securities and Exchange Commission. Manion set up a meeting between Markopolos and Grant Ward, the director of enforcement for the SEC in New England.

Markopolos spent hours sweating out his presentation. He prepared charts comparing Madoff's performance against the S&P. He analyzed the Broyhill fund record reflecting Madoff's performance over seven years. He compiled a blizzard of facts and formulas to prove that Madoff's investment returns were impossible to achieve. He readied his argument that the king of NASDAQ was running history's largest Ponzi scheme. Yet it may have been his personality that proved most memorable.

Ushered into a large room, Manion introduced Markopolos to Ward. Then the two SEC officials had a seat and let Markopolos make his case.

"In 25 minutes or less, I will prove one of three scenarios regarding Madoff's hedge fund operation," Markopolos began, according to his typed formal presentation.

One—They are incredibly talented and/or lucky and I'm an idiot for wasting your time.

Two—The returns are real but they are coming from some process other than the one being advertised, in which case an investigation is in order.

Three—The entire fund is nothing more than a Ponzi scheme.

His presentation was dotted with complex mathematical calculations and back-office jargon, with exhibits that only a fellow rocket scientist could love. ("Part A, a split-strike conversion is long 30–35 stocks that track the 100 stock OEX index, short out of the money [Delta < .5] OEX index call options.")

Amid the dense thicket of numbers and formulas were hints of his eccentricities. He railed against his own company, complaining against managers who threw Madoff's returns in his face. "My firm's marketing department has asked our investment department to duplicate Madoff's 'split-strike conversion' strategy in the hopes of duplicating their return stream," he said. "We know from bitter experience that this is impossible but they won't listen to my firm's investment professionals. . . . I would like to prove Madoff a fraud so that I don't have to listen to any more nonsense about split-strike conversions being a risk-free absolute return strategy."

Then he made a bald, premature pitch for the bounty money. "If there is a reward for uncovering fraud, I certainly deserve to be compensated," he said. "There is no way the SEC would uncover this on their own." He was demanding a reward before proving his case. "I have no hard evidence of fraud, just suspicions that things aren't what they seem inside of Madoff," he said.

But overall, his analysis of the Wall Street titan was far-sighted and should have been enough to ring alarm bells: he was alleging a fraud of up to $7 billion. "Combining the discrepancies I've noted . . . with the hearsay I've heard, seems to fit in with the patterns commonly found in Ponzi schemes," he concluded.

The blank stares that greeted him when he finished said it all. Ward was an attorney, not a CPA or an MBA, and no one without a background in Wall Street arcana could have fully followed the case Markopolos was laying out. Ward also didn't let on whether he considered Markopolos credible.

Markopolos never heard back from him. "All I can recall is that this fellow didn't understand much, if anything, that

I was presenting," he said. "He did not seem to have any formal background in finance or the capital markets. How he came to hold that senior of a position at the SEC was a real eye-opener and should have been my first clue that turning in this case to the SEC was a bad idea."

Those clues would mount in the coming years, but Markopolos mostly disregarded them. He was deeply immersed in the mind-set of the Army intelligence officer he once was, viewing his pursuit of a Wall Street charlatan as a patriotic mission, a defense of the American flag. It was a little over the top, and perhaps not even the entire story; by his own words, Markopolos seemed driven at least as much by money as by justice. But whatever his motivation, he was a driven man, determined to bring down Bernie Madoff. Fortunately for Madoff, no one was listening to him.

5

The Friendly Company

"What makes it fun for all of us is to walk into the office in the morning and see the rest of your family sitting there," Mark Madoff told a reporter in 2000. "To Bernie and Peter, that's what it's all about."

The older of Bernie and Ruth's two sons wasn't kidding. To work at Bernard L. Madoff Investment Securities was to either be a Madoff or be treated like one.

The finance business is among the most cut-throat of worlds, but Madoff's employees may have been the happiest and most coddled in the industry. The employees on the seventeenth, eighteenth, and nineteenth floors of the Lipstick Building lived a sheltered existence under Bernie's protective wing.

The millionaires of Palm Beach were too intimidated to speak to him, but inside his shop Bernie was entirely approachable. To most of his employees he was the amiable patriarch, walking the trading floor with a cardigan sweater over his shirt and tie, slapping backs and making small talk. When an employee fell behind on his mortgage, Bernie paid it off. When another got married, he paid for the honeymoon. The junior traders at the bottom of the payroll were making Wall Street's version of a paltry income, $150,000 a year, so Bernie gave them an extra two weeks' salary on top of their annual bonuses.

Every summer on the weekend following July 4th, he threw a stunningly expensive party for the entire office in Montauk, Long Island. Hundreds of hotel rooms were rented out for the company's interns, messengers, managers, technology experts, clerks, secretaries, and traders, plus their spouses and kids. The main event was a huge lobster dinner at the Montauk Yacht Club. There were no meetings, no speeches, and no obligations except to take advantage of the company's generosity.

All the traders had friends at places like Goldman Sachs and Merrill Lynch who arrived for work at six o'clock each morning, bleary-eyed and miserable. Madoff's traders, in contrast, were allowed to amble into work at eight. Their compensation was, if anything, higher than at other shops. Star traders were taking in $5 million a year.

Madoff Securities was an impressive place. The company was nestled inside a thirty-four-story Third Avenue skyscraper, an elliptically shaped building wrapped with ribbons of red granite and steel that resembled a tube of lipstick. New

Yorkers dubbed it the Lipstick Building when it opened, one year before Madoff moved there in 1987.

There were three legs to the company: market-making, proprietary trading, and investment advisory. The two trading divisions were spread across a large nineteenth floor trading room sprinkled with hundreds of blinking computer terminals. The proprietary traders played the stock market for the benefit of the company. The market makers brought buyers and sellers together.

Market-making and proprietary trading were widely respected entities, the shining face of Bernard L. Madoff Investment Securities. The two trading arms were kept separate from the third part of the firm, the secretive investment advisory business located on the seventeenth floor (the eighteenth floor housed administration and information technology). It was on seventeen that Madoff took in billions of dollars from friends, families, and financiers extending from France to Far Rockaway. It was where the vaunted split-strike conversion strategy was said to be managed.

Like much of the outside world, the traders on the nineteenth floor often wondered how the investment advisory business managed to do so well and why it was kept so secluded. Madoff seemed to ward off scrutiny simply by keeping his employees upstairs so happy. "Why would we question a good thing?" one asked.

Not surprisingly, a large number of the company's 150 employees worked for Madoff Securities for years; the newest employee of the IT department had been there for twelve years. The employees grew up together, spent time with each other outside work, knew each other's wives and kids, lived through one another's weddings, divorces, and illnesses. "It was the kind of company you knew you could work at for the rest of your life," said one. "No one ever got laid off. This was like a government job. You knew you could work here forever."

It was an unusual atmosphere. Elaine Solomon discovered that on her first day.

Had she been born into another family, the 48-year-old Englishwoman might have grown up to be an executive at a place like this. Raised in London and educated at its most elite boarding schools, she was the epitome of a proper British lady. She was attractive, bright, and sharp-witted, with an air of formality that commanded respect. But her parents had groomed her to be a wife and mother, and her education ended with a degree from a secretarial school. By 1997, after a failed marriage and stints as a flight attendant, bookkeeper, and legal secretary, she was living in New York and looking for her next job. Flipping through the classifieds, Solomon found a help-wanted ad seeking an executive assistant for Bernard L. Madoff Investment Securities. She sent in her résumé and shortly thereafter received a call from Bernie's niece, Shana Madoff, asking her to come in for an interview.

With her flawless manners and erudition, Solomon made an ideal executive assistant candidate, and Shana immediately liked her, not least because the candidate was also Jewish. "You're just like my mother!" Shana said. The job they offered her was as secretary to Peter Madoff, who was Shana's father, Bernie's younger brother, and the man in charge of daily operations. Solomon hadn't heard of the company or the Madoffs, but there was a casual feel to the place that she liked; it was a huge contrast to the uptight atmosphere of some of the other places she'd worked at. She was hired an hour after she walked through the doors, not something that usually happens at the headquarters of a billion-dollar corporation. But Madoff Securities was not your average billion-dollar corporation.

Solomon was given a tour of the office by the woman who would become her closest coworker, Bernie's secretary, Eleanor Squillari. "Elaine, this is Mark, Bernie's son. . . . This is Andy, his other son. . . . This is Charlie, Bernie's nephew. . . . This is Roger, also his nephew. . . . This is Ruth, Bernie's wife."

"Do you realize that everybody's related in this place?" Solomon asked.

"Oh, are they?" Squillari replied. "I guess they are." As Solomon would discover, Madoff's employees were so accustomed to the unusual that they barely recognized it anymore.

The new hire had hit on the fundamental oddity of the place. Trades approaching $2 billion a day were speeding through the circuits of Madoff's space-age trading system, but the company was managed like a small family restaurant.

The action revolved around Bernie's and Peter's nineteenth-floor offices down the hall from the trading floor.

The brothers worked across from each other, their offices separated by a glass conference room. Elaine Solomon and Eleanor Squillari sat a few feet away from each other outside their offices, serving as the family gatekeepers. Squillari, like Solomon, was a formidable woman, the sergeant of the executive suite and a fixture at the company since her hiring a decade earlier. The two women became fast friends.

The line between life and work was hopelessly blurred in the executive offices. Bernie's sons, Andy and Mark, who ran the trading operation, had offices on the same floor. One flight below, on the eighteenth floor, Ruth sometimes came in to go through Bernie's receipts and handle other bookkeeping chores in her office. Not far from her was Peter's daughter, Shana, the company's compliance lawyer. Andy's and Mark's desks sat on an elevated platform in the trading room, overlooking the market-making and proprietary trading areas. The brothers would often be joined by Bernie, Peter, or Shana, who'd gaze out on the action or take a conversation into a glass-enclosed conference room nearby known as "the Tank."

The family members were with each other constantly; at night they would dine together, fighting for the opportunity to sit next to Bernie. They all seemed to be plugged into the same issues, the same gossip, and the same problems that were being dealt with at any given time. "They didn't pee without telling each other," Solomon recalls.

There was an incestuous feel even among the nonfamily employees. Norman Levy, Madoff's old friend and a major client, had an office in the building. Sonny Cohn, Madoff's partner in his investment company Cohmad Securities, worked out of Madoff's offices; his daughter Marcia was his compliance officer.

Madoff's trusted feeders were in and out of the office all the time. Ezra Merkin, the brusque intellectual, would barge past Squillari and Solomon as if they didn't exist. Robert Jaffe, the Palm Beach socialite, would chat them up and lavish gifts on them at holiday time, often $400 gift certificates to the uptown wine store Sherry-Lehmann. There often seemed to be no boundaries to the place. Solomon would marvel at how all the Madoff men seemed to share the same bad habit of walking out of the executive men's room zipping up their flies.

No one was less aware of his boundaries than Bernie.

The boss was polite to Elaine when she first began, but the longer she worked there, the more liberties he took. He asked questions about her personal life and made crude jokes to Eleanor and her. "You know it excites you," he told Eleanor once as he zipped his fly coming out of the bathroom.

He liked to make the kind of sarcastic wisecracks that made people anxious. "I made you and I can break you," he'd tell employees. He could be so obnoxious that his assistants tried to just tune him out. When a pretty young staffer once walked into the office, Madoff glanced at Squillari and asked, "Do you remember when you used to look like that?" He smirked and tried to give his secretary a pat on the behind.

Squillari had been Madoff's assistant for over ten years and wasn't always willing to suffer his indignities. When Madoff asked her to have a messenger pick up his watch from Cartier, she pronounced the name without its French ending, prompting Madoff to teach her a lesson. "It's Car-tee-aay," he said. "Don't correct my English!" she snapped back.

When he was angry or frustrated, he would hurl horrible insults at his secretaries, Eleanor Squillari recalled. "You look terrible!" "You're an idiot!" "You're stupid!" This was not the Bernie Madoff seen by the rest of the world, the mysterious figure who barely uttered a word in public and refused to network on the golf course. In his inner sanctum, he could be gregarious, coarse, generous, gentle, rude, and sometimes vicious.

But the people who worked for him nevertheless considered him a father figure, always there for you when you needed him most. In the Madoff constellation Bernie was the sun around whom everyone else revolved. His staff, friends, and family all seemed to exist to keep him happy and meet his needs. "Whatever Bernie said, you did," Solomon recalled. "If it was Thursday and Bernie said it was Tuesday, then you said 'You're right.' You didn't argue with Bernie."

No one was more focused on his needs than Ruth. The love and care of Bernie was her full-time job, far more so than the bookkeeping she attended to every once in a while down on the eighteenth floor. Each morning, she'd call Eleanor or Elaine with a rundown of Bernie's needs. "Bernie wants a massage," she'd say before a trip. "Can you make sure he gets a massage every day at the hotel?" Ruth called him "Darling" and gazed at him with the deep affection she likely displayed as a teenager in Laurelton. Bernie, always reserved and unemotional, clearly considered Ruth his closest confidante and showed his faith in her by entrusting her with the company's books.

Squillari later wrote in *Vanity Fair* that there were two Ruths: one a sharp and attractive woman with a beautiful figure and stunning taste, the other an "aging blonde who seemed to wish she were taller, younger, prettier." In her less secure moods, she would lose patience with Bernie if he hit her with one of his hurtful remarks. "Go fuck yourself," she'd tell him.

Her eyes were always following her husband. It wasn't paranoia; Squillari had once caught Madoff poring through ads for escort services, and she found a list of masseuses in his address book. "If you ever lose your address book and somebody finds it, they're going to think you're a pervert," she told him. She and Solomon would note that he often left the office in the afternoons and returned smiling later in the day.

The atmosphere in the office was set almost completely by the fluctuating emotions of a family and its mercurial patriarch. "There was nothing corporate about it," Solomon said. "Nothing."

In the 1980s, Bernie Madoff hit on a jackpot solution for luring business to his market-making operation. It was as simple as it was inelegant: he paid for it.

Until then, investment houses such as Morgan Stanley and Merrill Lynch paid a fee to the New York Stock Exchange for the privilege of trading on the world's most important exchange. Madoff reversed the relationship: he offered to pay companies to trade through him. It was basically a legal kickback to lure firms to throw him work. The NYSE howled that Madoff was bribing his customers and enticing them to abrogate their responsibility to their clients. But Washington regulators liked the fact that Madoff was infusing competition into Wall Street and refused to intervene.

The business that proceeded to pour into Madoff Securities required a far larger operation, and Peter designed a cutting-edge trading operation that was the envy of Wall Street.

Traders on the floor of the New York Stock Exchange were still shouting and waving paper forms in their hands when the Madoff trading system came online. Trades at Madoff Securities were fully computerized and processed far faster than at the NYSE. Stock prices were captured in real time, often to the profit of both Madoff and his customers.

The concept had been conceived by Bernie, but it was realized by Peter. Bernie had lived in his younger brother's shadow as a child, struggling to keep up in class while Peter became known as one of the brightest kids in Laurelton. Bernie went to second-tier colleges and dropped out of law school after a year. Peter graduated from Queens College and Fordham Law. It was one of those unexpected turnabouts in life that the Madoff brother who was the star of the family ended up subordinated to his brother the struggler. As an adult, Peter continued to strike people as the more percep-

tive and forward-thinking brother, with a mind able to design one of Wall Street's most advanced trading systems. But
Bernie was the Wall Street titan and the one with his name
on the company's front door. Peter worked for him. Bernie's
license plate was "MADF." Peter's was "MADF2."

Employees would sometimes wince as Bernie talked down
to his younger brother. William Nasi, a longtime messenger
for the company, says he heard Madoff screaming at Peter
early one morning, "Until your name is on that door you
keep your fucking mouth shut!"

As director of day-to-day operations, Peter was the one to
crack the whip with employees while Bernie played the *ki-
bitzer.* "You have nothing to worry about from me," he told a
new hire with a mischievous smile. "Peter's the one you have
to worry about." But underneath the surface it was Peter who
was kinder and more thoughtful. He was a busy man, but
he set aside time to send a messenger to White Castle, order
hundreds of burgers, and hold an office burger-eating contest
(which Peter won). His employees thought the world of him.

Bernie had deep love for Peter and sometimes seemed
protective of him despite his propensity for demeaning him.
He trusted him enough to run the trading arms of his company and valued him so much that the brothers flew in separate planes just to ensure the company could never lose both
of them at once.

Had they both passed away, the next generation was already installed into management. How prepared they were
to take over the company was debatable.

Bernie's sons, Mark and Andy, were the princes of the palace, with an air of entitlement that came with the bloodline. In the Roslyn High School yearbook, Andy, the younger son, held a beer bottle in his photos and hinted at being born to the throne: "FUTURE PARTNER MADF—MOM, DAD—UR THE BEST "U'LL ALL SEE—I'M THE 1!" Andy was the more cerebral of "the boys," as the family called them, handsome and blond with a personality of "dry toast," in the words of a Madoff portfolio manager. He could be cold to his colleagues and indifferent to staff—"He acted like British royalty," said Errol Sibbley, his driver—and he seemed to detect that his father trusted Mark more. "Why is he always looking over my shoulder?" Sibbley overheard Andy complain to his brother.

Mark was charismatic, relaxed, and far better liked around the floor, where he cracked jokes with traders and called them by nicknames he invented. He was the apple of Bernie and Ruth's eyes—she called him "Marky Mark"—and spent far more time in his father's office than did Andy; some thought he was being groomed as the company's future president.

To a lot of the traders who worked under them, neither of the brothers seemed especially motivated. Like his father, Mark loved the trappings of wealth, and wasn't nearly as reluctant to flaunt it. He hung out at the Core Club a few blocks down, whose $50,000 membership fee and $15,000 annual dues bought him the opportunity to rub elbows with Bill Clinton and Harvey Weinstein. Both brothers were obsessed with fly-fishing, and were often off on exotic fishing expeditions. They also invested in companies catering to the sport. More than a few people on their staff saw them as self-involved and entitled.

"The market would open at nine-thirty in the morning and it would be dropping five percent and we'd all be in horror and shock," the portfolio manager recalled, "and Mark would be in his office on the J.Crew website, shopping."

"They had no fire in the belly," he said. "Zero."

Bernie often seemed disappointed in his sons, and demeaned them in cruel ways. He belittled their accomplishments in front of the staff and implied that neither measured up. It led to a good amount of tension, with the boys resenting their father for his behavior—and their mother for reflexively siding with him.

They were forever trying to prove themselves to Bernie. On the morning of March 2, 2000, Andy decided to plunge into trading himself. It was the day that Palm, the company that gave America the Palm Pilot, went public. "What do you think?" Andy asked some traders. "These things tend to skyrocket. If you're there and accumulate enough, you can turn it around for a huge profit."

Every stock trade requires a buyer and a seller. Market makers serve both roles. They set a price at which they'll buy a stock and the price at which they'll sell it, thus making a market for the stock. Their profit comes on the spread between what they buy and sell it for.

Andy decided to get into "the box" on Palm's initial public offering, Wall Street-speak for diving in and making a market. It was as simple as going to a NASDAQ workstation, clicking the mouse a few times, and setting up a continuous two-sided quote: buy and sell.

When trading opened at 9:30, the eyes of the trading floor were on Andy. It did not go well. He immediately outbid the market by about $35 per share, only to watch in horror as the value of the stock he'd bought plummeted $70 to $95 per share. With the pressure mounting, a systems glitch kept him bidding at the staggeringly high price of $165. The trading floor fell into a panic.

Bernie burst out of his office. "What the hell is going on?" he yelled. But the nightmare was in full bloom. His son's decision to dabble in the market cost the company about $8 million. The stress led Mark to smash his Palm Pilot on his desk.

The boys were not the only members of the next generation running major divisions of the company. Peter's daughter, Shana, the company's compliance lawyer, was smart and well-educated, and she had the reputation of being a tough, bright attorney. Like her cousins, she also had a taste for the good life that being a Madoff afforded. She was a fashionista, the proud owner of dozens of pairs of Manolo shoes. She was so fashion-conscious that *New York* magazine featured her as one of three extreme brand loyalists in a 2004 story: "Shana Madoff. Obsession: Narciso Rodriguez. Extravagance: Black." She left the tough buying decisions to Rodriguez's salespeople, she said, asking them for a shipment of clothes every season and to charge her for what she didn't return. "I just don't have time to shop," she told the magazine. "I could be doing so many other things that are so much more productive. And the salespeople are around the clothes all day. They know them much better than I do."

There was no one in the company capable of saying no to the Madoffs. The family members were often on vacation and came and went as they pleased. Weeks would sometimes pass without a sighting of Bernie and Ruth; they'd be off at one of their many homes, in Palm Beach, Montauk, or Cap d'Antibes in the South of France. These habits spread to the other people in the executive suite. Solomon and Squillari spent hours wasting time; Eleanor made jewelry at her desk, while Elaine played card games on her computer. "My brilliant secretaries," Bernie would announce sarcastically.

Yet the image of family management was a huge selling point. The firm's website boasted about it as proof of the love and care put into the operation. It was the reason one of Ruth and Bernie's closest friends invested his life savings with Madoff, worth tens of millions of dollars. "One of the safety nets I always thought existed was that Bernie could never do something to hurt his family," he said.

Life for a Madoff family member was fun, Mark told a reporter. But for all the freedom and swagger inside the executive suite, family members and staffers also had to suffer through the boss's stranger side.

Madoff was a captive to his compulsions. He was so obsessed with neatness that he'd get down on his hands and knees in the reception area to straighten out the Persian rug leading to the elevator. A staffer once found him vacuuming his floor at 7:30 in the morning. He was constantly adjusting the blinds in the trading room, grumbling about the disorder as his neurotic blink kicked in.

His focus on appearance bordered on mania. The office was painted and furnished in just two colors: black and gray. The carpeting was gray; the textured wallpaper was gray; the door frames were black. For a while, the company banned picture frames from all desks. Then the rule was loosened to allow only black or gray ones.

The Madoffs enforced Bernie's orders with fear in their voices. Peter once spotted Ken Hutchinson, a computer programmer, using a blue pen. "Where'd you get that?" Peter demanded. "We don't use blue ink here." It didn't match the decor. A trader recalled Mark noticing potato chips on his desk one day. "I don't care how much money you make," Mark told him. "Bernie will fire you for this." His father would harangue employees about sloppy desks and cluttered work areas. "What's all the crap on your desk?" Bernie would demand. "Clean it up." Sometimes he'd grow so frustrated with Elaine Solomon's desk that he'd clean it up himself.

He was beyond meticulous when it came to clothing. His secretaries had an entire cabinet full of cashmere sweater vests for when he got cold in the office. When Bernie heard rumors that the shoe manufacturer Belgian Shoes was going out of business, he bought fifty-five pairs for himself.

His extreme behavior was even more apparent in his work habits. He was enormously secretive and was consumed with knowing every bit of information that flowed into his offices. As fond as he was of Squillari, he never trusted her with anything confidential; he did all his own filing and took his papers with him in a briefcase wherever he went. "I know you're honest," he told Solomon one day, "but why Peter lets you know all about his finances, I don't understand it."

He took every phone call and insisted he be told the names of everyone who called, even if they hadn't left messages. Each day at 5:30 p.m., after his secretaries had gone home, the receptionist had to place a call to Bernie even if there was nothing to report. "Hi, it's Jean, Bernie. No messages, no faxes." When he was away from his office, his secretaries were under strict orders to conference him into every call, regardless of whether he was in the building or even in the country. The Madoffs were movie fanatics and would clear out of the office on Tuesday afternoons to attend a weekly movie club. When Solomon patched in a call for Bernie, he'd sometimes answer with a whisper. "Where are you?" she'd ask. "In the theater," he'd reply. "Let me go into the lobby."

The staff would speculate that Bernie was a man with something to hide. But more often than not they attributed his behavior to his eccentricities. It was all just part of what made Bernie the genius that he was.

On a November day in 2003, a childhood friend gave Madoff a call at the office. Donny Rosenzweig may have been the last boy to kiss young Ruthie Alpern before she fell in love with Bernie in Laurelton. She lived down the street from Donny, and they started to go to movies together at the Itch. She became his first girlfriend when they were 13.

Many in the neighborhood knew one another, but Donny had never gotten to know Bernie until Donny's parents hired him to lay down their sprinkler system. Donny would watch Bernie crawl through the grass of his parents' lawn, digging ditches on his hands and knees. Like so many others, he was impressed with his work ethic.

Eventually, Ruth and Donny drifted apart and she moved on to dating Bernie. Donny would occasionally wave to the couple outside Ruth's house when he visited his grandparents on her block.

It would be another fifty years before Rosenzweig had another conversation with Madoff. He had built up his own textile manufacturing business and sold it for several million dollars. Now he was looking to invest the money.

Everyone from Bernie's era in Laurelton had kept track of him and buzzed about all the money he'd made on Wall Street. Rosenzweig figured he had nothing to lose by picking up the phone and seeing if Madoff remembered him. To his surprise, Bernie took his call. (He was unaware that Bernie took all his calls.) And Bernie did remember him. "Donny, how are ya?" he said, seeming genuinely happy to hear from him. They talked about old times, and then Donny hit him with a question. He'd found a fund of funds—a kind of mutual fund comprised of hedge funds—that was earning about 7 percent a year. What did Bernie think about Donny placing his money with it? Bernie thought for a moment.

"Let me tell you what we're doing here," he said. His asset management fund for private investors was currently returning 12 to 13 percent, he said.

But it was a giant tease. Rosenzweig had been following Madoff's success from a distance for decades and had friends who'd made a lot of money with him. But when he popped the question about getting into his fund, Madoff declined. "I can't," he said. "It's a closed fund." At that, Bernie told him there was someone in his office who wanted to speak with him. He handed the phone to Ruth.

The two former flames chatted excitedly, mostly about old times, until she stopped him in mid-sentence. "Bernie wants to talk to you again," she said.

Bernie took the phone back. "You know what," he told Rosenzweig, "we've known you a long time, and I'll open it up for you to come in. . . . But can you meet the minimum?" Bernie cited $2 million. "I can meet that," Rosenzweig said. "Do you mind if I bring others in with me, some family?" Not at all, Madoff said.

It was a happy experience, the first of many. The Rosenzweigs took Bernie and Ruth out to dinner, and they spent the night mostly talking about the Madoff family. Bernie talked earnestly of not wanting to overshadow his sons so that they wouldn't feel successful in their own right, his demeaning behavior notwithstanding.

In the coming months, Bernie invited Donny up to the nineteenth floor to tour his offices. "They were very neat," Rosenzweig recalled. They talked about Bernie's split-strike conversion strategy, how he'd buy twenty-five stocks from the S&P 100 and hedge them with options. "The key is making sure the stocks represent closely what the S&P does," Madoff said, flying in the face of what Harry Markopolos had discovered. "Then the options you buy will do for you what they do for you."

"Don't try to match the stocks I buy," he warned Rosenzweig. "We do it totally differently than the way you'd do it on your own. You'd get hurt if you tried." Rosenzweig was thrust under Madoff's spell. As the years passed, he poured more of his money into his old friend's investment fund, along with the savings of his family members.

It would be five years before Rosenzweig learned that Madoff never invested his money at all. As the two old friends continued to socialize, Bernie was stealing it all to pay off other investors and support his luxurious lifestyle. All the while, he never displayed anything but joy to be helping out an old friend. "I really believed he was doing me a personal favor," Rosenzweig said.

Like most of the world, Rosenzweig viewed Madoff as a *mensch*, generous to his staff, kind to his old friends, wise with people's money. Only those closest to him on the nineteenth floor saw glimpses of a more troubled side. Yet even they would be shocked when they learned the secret that had driven him to behave so oddly through the years.

The only employees who might have known what was bugging Bernie Madoff were working two floors down.

6

The Seventeenth Floor

It was one of Clive Brown's morning rituals. At 10:30, the Jamaican-born company driver would walk into a seventeenth-floor office in the Lipstick Building and wait for Norman Levy's check.

Brown was a courteous, impeccably dressed, middle-aged man with a deep loyalty to the Madoffs. Bernie had been a gentleman to him, the kind of boss who would catch a taxi home in the rain rather than make Brown wait behind the wheel into the evening. "Go home to your family," he'd tell him. When Bernie hired Brown in 1987, he put his hand on his shoulder and gave him a fatherly smile. "You'll never have to worry about your job here," he said. "As long as you don't steal."

The seventeenth floor was a mystery to Madoff's employees on the trading floor. Regardless of how important they were or how many millions of dollars they were making, their electronic card keys wouldn't unlock its doors.

But the card keys held by the company's four drivers opened every door.

Each morning, Brown entered the seventeenth-floor office area housing Madoff's investment advisory business and waited quietly for Levy's check. The office suite was the bureaucratic nerve center of Madoff's investment operation, the department that handled the accounts of Carl Shapiro, Elie Wiesel, Donny Rosenzweig, Stephen Richards, and thousands of other private clients. It was where the split-strike conversion strategy was presumed to be carried out, where the money from the elderly residents of the Sunny Oaks and the members of the Palm Beach Country Club flowed in, where the monthly and quarterly statements showing consistent profits were printed and mailed out to overjoyed customers.

The area had a dreary sweatshop feel to it. Bernie's former secretary, a squat disciplinarian named Annette Bongiorno, often yelled at her staff of young female clerical workers, who joylessly punched data into computer terminals all day. Bongiorno regularly kept them working late, forbade them from making personal phone calls, and eventually disconnected their email service. "She was a terrible slave-driver," an employee recalled. "Everyone who worked under her hated her."

The employees themselves were shockingly nasty to clients. Legions of angry customers regularly called upstairs to Elaine Solomon in the executive suite to complain that they'd been yelled at or mistreated by the girls downstairs. But Solomon could only offer them her sympathy; whenever she called downstairs, they'd bite her head off too.

Annette and other managers shared a table near the front door, none of them particularly welcoming to visitors who appeared before them. "You'd go there and people would be looking at you saying, 'Do you have business here or not?' " recalled a support staff employee.

Each morning, as Clive Brown waited, someone in the seventeenth floor office would prepare Norman Levy's check for him. The routine was so old there was no need to tell Brown what to do with it.

Like Carl Shapiro, Levy was an enormously wealthy man who had fallen under Madoff's spell decades earlier, when Bernie was just beginning. He was large and garrulous, almost six and a half feet tall, and fond of boasting about his humble beginnings in the Bronx. He had fought his way into business "back in the days when Jews couldn't sell real estate," he told Brown. In the decades that followed, he became a giant of the New York City real estate industry, with an ownership stake in dozens of skyscrapers, including the famed Seagram Building.

Madoff's relationship with Levy transcended virtually all others in his life aside from his family. Madoff's father died at a relatively young age, and Bernie would often speak of Levy as a son would of a father. There was no line separating business and friendship between the two, which was not atypical of Madoff's way of operating. His mentors tended also to be his benefactors, and Levy invested hundreds of millions of dollars through Madoff and raised millions more for him from his fellow business magnates. He called Bernie at least once a day ("Hey, big guy"), often with the name of a stock he wanted to buy, as though Bernie were a junior broker. Eventually, Levy left the JPMorgan Chase Tower and moved into the Lipstick Building to be closer to his protégé. The two men and their wives dined out together, celebrated one another's birthdays, vacationed together, and worked on the same Jewish charities. Madoff treated Levy like royalty. He had Clive Brown chauffeur him around the city, planned Norman's vacations, even helped oversee the construction of Levy's 127-foot yacht. Levy thought the world of him. "If there's one honorable person," he told his son Francis, "it's Bernie."

Brown would examine each check for Levy that he picked up on the seventeenth floor each morning. They were usually for millions of dollars. Every now and then they'd be for tens of millions.

The driver didn't need a car where he was going. He walked the check over to JPMorgan Chase's headquarters at 270 Park Avenue, eight blocks from the Lipstick Building, and either handed it to Levy's secretary upstairs or deposited it in the bank downstairs. Then he headed back to the Lipstick Building and back to his work as a chauffeur.

Brown always assumed the checks were the profits from Levy's stock transactions. But there was no rational reason why someone would cash in his returns every day, much less have them hand-delivered to his office. Even if there were, Levy's trading profits wouldn't have been nearly that large. Perhaps somewhere there was a perfectly legitimate explanation for these deposits. But it's equally plausible that Bernie was using his friend's bank account to park stolen money.

Chase also housed Madoff's main investment advisory account, where most of his investors' billions resided. He and Levy were somehow able to go about their daily banking routines completely unnoticed by the bank's internal compliance systems.

Like everything that was illicit in Madoff's company, the checks were emanating from his criminal operation on the seventeenth floor.

It may or may not have required a rocket scientist to realize there was a problem with Bernie Madoff's investment business. But it certainly didn't take one to realize something was strange about the man he chose to run it.

Frank DiPascali walked around like a neighborhood tough. He was small and wiry, with a deep Florida tan and short, groomed hair, a familiar look for a middle-aged man from the Italian community of Howard Beach, Queens. The company enforced a strict dress code, but DiPascali traveled the halls in jeans and a T-shirt, looking more like the telephone repair man than a high-level official. Some employees thought he looked like a loan shark. Most tried to stay out of his way.

He had come to the company a year after graduating from

Archbishop Molloy High School in 1974, his brief college career collapsing just a few months after it began. In the years that followed, he rose to become Madoff's "chief financial officer," as he told some, "director of options trading," as he told others. Whatever the title, DiPascali was a curious choice to oversee one of the world's largest investment funds.

He clearly had the run of the place. If he walked onto the trading floor, people would defer to him. If he ordered a new copy machine, no one in the IT department asked if he had budget approval—they just sent it over. If he had a problem with his computer, technicians would drop what they were doing to help him. The fact that he dressed in jeans in a shop that strictly forbade them was all you needed to know about his standing at Madoff Securities.

In an office filled with Ivy Leaguers and MBAs, he looked like he'd gotten off on the wrong floor. His contempt for the yuppie culture at the office became abundantly clear when he went looking for a summer intern in 1990.

Many of the students who descended on the company each summer were the sons and daughters of Bernie's rich friends, kids who had little interest in or talent for securities trading. Some were students from prestigious colleges headed for business school. DiPascali had little use for any of them. Bypassing stacks of internship applications, he decided to pay a visit to his old high school in Queens.

Archbishop Molloy is the kind of institution one never totally leaves. It is a place beloved by generations of middle-class Queens residents for its Catholic values, solid academics, and championship sports teams. DiPascali entered a school building graced by a large bronze statue of the Virgin Mary and walked through hallways filled with girls in prim gray skirts and boys in white shirts and ties. Three decades earlier, he'd been one of them.

He headed to the guidance counselor's office and asked the brothers there to find him a promising student. They didn't need to think long. They produced a lanky senior named Ken Hutchinson, a technical whiz who had puttered around the school's computers since fifth grade and basically taught himself to program. He was modest, respectful of authority, and serious about his future. He didn't come from privilege; he was the product of a middle-class family from Forest Hills. His dad had worked for the government for forty years. He was an ideal candidate.

DiPascali met him for a brief interview—"He basically wanted to make sure I didn't have purple hair or something," Hutchinson said—and offered him the internship. Hutchinson had been planning to spend the summer stocking shelves in a grocery store. The offer was a godsend.

His new life at Bernard Madoff Securities began a few months later on the old eighteenth-floor trading room, copying documents and rushing through the aisles to deliver breaking news from the Reuters tape machine to traders at their terminals. There were supervisors there who treated him "like garbage," Hutchinson said, but from his first day at Madoff he seemed to occupy a special place in DiPascali's heart. Each week, DiPascali would motion for Hutchinson to join him in his office. He'd pull out a pad and a pen and proceed to teach him the business, from options strategies to arbitrage—"the art of trading," Hutchinson said. "This is how the market works," DiPascali would begin. He sprinkled his lessons with disparaging comments about the self-important young professionals outside on the trading floor, and lumped the Madoff sons in with them. "Frank didn't like these poor little rich kids," Hutchinson said. "He was a self-made man."

When the intern started at NYU in the fall of 1990, DiPascali got him a part-time job in the IT department, computers being Hutchinson's first love. By then, he considered DiPascali a mentor. "If you looked at him, you'd think he was an electrician," Hutchinson recalled. "But his looks were deceiving. He was very smart."

Business was booming at the time, and the company expanded its operation. The market-making and proprietary trading departments moved up to a new trading floor on nineteen and the investment advisory group moved down to seventeen. DiPascali moved downstairs with it.

But the ground soon shifted from underneath the Madoff empire. By the mid-1990s, NASDAQ market makers were feeling increasingly embattled as the SEC and Justice Department brought an end to the industry's questionable business practices. Reprimands started flying against Wall Street companies for selling stocks at inflated prices. Market makers traditionally followed a silent code to update the price of stocks in 25-cent increments, which translated into huge profits but cost investors serious money. Madoff was forced to cut the increments in half, to 12.5 cents. Then in 2000, the SEC forced exchanges to list stock prices in decimals rather than fractions, which brought trading profits even lower. Profits for firms like Madoff's fell by more than half a billion dollars between 2000 and 2005.

The profitability of Bernard L. Madoff Investment Securities was starting to decline. In a 1999 press release, the firm boasted that it was a "leading market-maker" in more than two hundred NASDAQ stocks. By 2005, it was handling only half of 1 percent of NASDAQ trading volume. Goldman Sachs was handling ten times Madoff's volume.

As late as 2002, Charles Schwab and Goldman Sachs were said to be interested in buying Madoff's market-making arm. Bernie winced at the thought of selling it, even for a billion dollars. "I couldn't work for anybody," he said. "My whole family works here." The process of a sale would also by necessity reveal everything about each part of the company, from bookkeeping to trading strategies. Considering what was going on downstairs on the seventeenth floor, it wouldn't have been a good idea.

It increasingly fell on the investment advisory business to subsidize the trading operation. Madoff had to shift millions of dollars from the seventeenth floor to keep the nineteenth floor running. His flagship business, market making, was headed for deep trouble, and the only way to keep things rolling was to raise more money for his sweatshop downstairs.

● ● ●

Ken Hutchinson was hired full time in 1994 and rose quickly through the ranks of the IT department to become its lead database developer. Frank DiPascali was spending most of his time downstairs on seventeen and had become less of a presence in Hutchinson's life. The protégé began to see his mentor more objectively.

The IT guys worked one flight up on the eighteenth floor, which was considered unglamorous back-office territory. Workers in Hutchinson's area spent their days programming software and fixing the nuts and bolts of computer terminals. But they were in a unique position to observe the workings of the company. People who fix computers are often treated like maids at hotels: they're allowed in everywhere, and no one seems to notice them.

Going down to seventeen was "an event," Hutchinson said. Each time his colleagues returned, they would gossip about the odd things they'd seen. While Madoff made sure that every window blind on eighteen and nineteen was straight and picture frames didn't clash with the carpeting, the investment advisory offices were sloppy and cluttered. Old broken computer terminals gathered dust on the shelves. Documents and files were jammed into desks and thrown on shelves.

DiPascali was increasingly a creature of the seventeenth floor, fading from the trading floor he once helped operate. He and Bongiorno ruled the roost down there. They had been next-door neighbors in Howard Beach back in the late 1960s, when they came to work for Madoff; she had brought him into the company. Both had come a long way since then; each owned a pair of Mercedes and lived in multimillion-dollar homes. They were on one another's wavelength, street-smart, efficient, and short-tempered.

DiPascali, the senior manager of the two, had an office behind the seventeenth-floor workers. Just as his attire violated the company's dress code, his office violated its decor; it was the only room in the entire company without glass walls. When he closed his door, no one could tell what was going on in there.

"You didn't dare go in there or knock on that door unless the building was burning down," Hutchinson said. "He'd have no problem saying 'Get the hell out of here—you don't belong here!'" Hutchinson would often see Bernie go in and out of DiPascali's office. Bernie's family members rarely, if ever, appeared there.

It eventually became clear to Hutchinson why DiPascali had so much clout at the company and could violate so many of its rules. It was his proximity to Bernie.

Day after day, he watched DiPascali and Madoff retreat into private meetings. If they weren't in DiPascali's office on seventeen or Bernie's office on nineteen, they were in the Tank, the large, soundproofed, glass-enclosed conference room on the trading floor. When they walked into the Tank together, Bernie would flip a switch and send inert gas shooting between the double-paned glass walls, turning them instantly dark. Hutchinson never saw Peter, Andy, or Mark with them, which was unusual, considering how the family ran the place as a tribe. Whatever was said between the two men stayed with them.

Employees were amazed to watch DiPascali talk back to the boss; he was the only employee who could get away with it without ending up on the street. Driving Madoff through Manhattan one day, Clive Brown overheard him admonish DiPascali over the speakerphone after getting a complaint from a client about DiPascali's abusive phone manners. "You need to be calmer with her," Madoff said.

"Shit no," DiPascali replied. "She's a nobody."

The dispute later exploded into a full-blown, obscenity-filled screaming match on the seventeenth floor. Brown accidentally walked right into it, and tiptoed out just as fast.

Now and then, Hutchinson and DiPascali would see each other in the hallways and have a warm chat about how things were going for the former intern. But the protégé more often than not watched his old mentor from a distance. And what DiPascali was doing seemed increasingly strange to him.

The IBM AS/400 was born in 1988 and promoted by the company as the cutting edge of mainframe computing. It was sturdy as a tank and better protected against security attacks than any IBM office product in history, a quality that came in handy for an ultra-secret operation. But by 2000, many in the industry considered it an antique. It was a big, heavy metal box connected to a network of "dumb" terminals, pre-PC computers with no memory or hard drives. Financial houses all over Wall Street had long replaced them with modern-day servers and PCs. To the computer guys on eighteen, it bordered on embarrassing that one of the most technologically advanced financial houses in New York was still using this vintage monster. The computer system on the trading floor was the envy of Wall Street, but the company also retained two of the old clunkers, one for clearing the nineteenth-floor trades at the end of each day, and the other running the entire investment advisory operation downstairs.

DiPascali assigned two IT employees, George Perez and Jerry O'Hara, to work exclusively on the AS/400 on seventeen, with strict orders that no else was to touch the unit. There were people back at the IT department who were more experienced with mainframe computers, but it didn't matter; only Perez and O'Hara were authorized to come near the thing. Before long, he had them moved off the eighteenth floor altogether and relocated downstairs.

"A lot of people thought it was weird," Hutchinson said. But DiPascali convinced Hutchinson and his colleagues that the move was made to prevent the outside world from learning the split-strike conversion strategy's algorithmic formula, the company's secret sauce.

He came to wonder why visitors were never shown the investment advisory offices. Bernie was constantly giving tours though his gleaming trading floor to prospective investors. Traders listened as Madoff seemed to be representing the trading floor as the hub of his investment management operation. The corporate dungeon below didn't seem to exist.

The traders had other questions. No one on the vast nineteenth-floor trading floor had ever been asked to buy shares on behalf of Madoff's private investment clients. The employees on seventeen were supposedly initiating those trades separately and for some reason having them executed in Europe. It barely made sense, but everyone seemed to take it as a given. Andy and Mark bristled that their father didn't trust them with it.

New hires would join the IT staff and wonder all over again about the company's way of doing things. An IT project manager named Bob McMahon was brought in by a well-meaning technical manager who wanted to improve communications between the company's various computer systems. But he immediately ran into resistance. "Nobody could give me a straight answer why there were these disparate technology systems there," he said. The AS/400 surprised him most. "Why do you still maintain this gorilla?" he asked colleagues. "At some point, this thing is going to blow up and come to a screeching halt." He got sympathetic looks but no answers.

So many things about asset management puzzled him. "All I saw on seventeen was antiquated equipment—dot-matrix printers, things I hadn't seen in years," he said. "Everything just looked old, like walking into a Smithsonian world of the 1970s and 1980s." But his colleagues just seemed to accept it. "Someone just rolled their eyes and said 'They don't throw anything out there,'" he said. To overhaul the system on seventeen would require a team to examine the AS/400's trading data going back years. No one running the company seemed interested in considering the idea.

McMahon spent months complaining to his superiors that employees were throwing up one brick wall after another. Eventually the company let him go. "They told me I wasn't technology savvy enough," he said.

There was abundant evidence that something was wrong on the seventeenth floor. In the best of circumstances, employees like Hutchinson and McMahon would have recognized it for what it was, brought their suspicions to an authority that was willing to listen, and derailed a criminal operation that was swindling thousands of people out of billions of dollars. But Bernie Madoff was born under a lucky star. Evidence of his scam was all around, but no one inside his firm connected the dots. "It was the furthest thing from your mind," Hutchinson said later. "When the weird things came up, they were explained away. It was their esoteric way of doing things."

He believed in Frank DiPascali nearly to the end. "I never lost respect for him," Hutchinson recalled. "The entire time, I thought he was just making money for the company and doing his thing."

The heart of Bernie Madoff's billion-dollar Ponzi scheme was a sterile, cluttered office where a handful of unhappy clerical staffers punched stock trade information into old computer terminals. Madoff or his lieutenants were checking the stock returns from previous days and weeks and instructing the clerks to enter transactions that were based on old results. The computer system would apply the same formula to each client's account, the only difference being the number of shares each of them owned.

It was the simplest of crimes. There was no manipulation of the markets, no insider trading, nothing that required complex skills. The clerks who worked on seventeen were simply writing fiction, duping investors into thinking they were making money when in reality the money didn't exist. Most had no education in finance and couldn't tell you the difference between a put and a call. But there they were each day, inputting fictitious numbers that were being fed to them by their supervisors and spitting out results proclaiming to thrilled investors that Madoff had once again beaten the market.

It didn't take much ingenuity—any corrupt stockbroker could have done the same thing. Bernie Madoff was a far cry from the investing genius that so enchanted his legions of investors. His virtuoso abilities were pure invention. He was still the kid from Laurelton, masking his limited talents with hard work and street smarts. He managed to fool people not just about his investment business, but about himself.

7

The Black Box

In 2000, the Sunny Oaks Resort in the Catskills died its inevitable death. The Borsht Belt scene had long since faded, and the hotel's elderly Jewish clientele was dying off one by one. The resort's owners, Ted and Cynthia Arenson, struggled to fill the growing vacancies, charging just $50 a night for a room, with three meals a day and free folk dancing thrown in. But there were few takers. "Young people didn't want to stay with us," Cynthia said. "They didn't like all the eighty-year-olds."

But for the survivors of the Sunny Oaks and their children, the good times didn't end when the hotel closed its dilapidated doors for the last time.

One of its regulars, Saul Alpern, Bernie Madoff's father-in-law, had died the previous year after a slow descent into senility. He had long since handed over his accounting business to Frank Avellino and Michael Bienes but continued to serve as an operative for Madoff, signing up investors over gin rummy games on the Sunny Oaks pool deck. Dozens of residents had enthusiastically given Madoff their life savings to invest, and the returns were providing them with wonderful retirements. Many moved to high-end assisted-living communities. Some put their grandchildren through college and law school. Others helped their kids buy new houses or cars.

The Arensons had money with Madoff too; Cynthia's mother had left her a Madoff account worth about $300,000 when she died, and it became a gift that kept on giving for the couple. The profits allowed them to bulldoze the Sunny Oaks and build a beautiful two-story vacation home for themselves, with sixteen-foot ceilings and an eight-foot-deep swimming pool.

Saul had guaranteed his fellow Sunny Oaks residents 20 percent annual returns, a number that only gradually slid over time, and Madoff delivered year after year. No one can recall anyone questioning how someone could guarantee them such a windfall.

The crowd at the Sunny Oaks wasn't alone in its good fortune. Hundreds of Bernie Madoff's investors were retiring early, moving into bigger houses, and traveling the world on the money they were making from their investments.

Stephen Richards counted himself a blessed man for having met Madoff at Saul Alpern's house in Laurelton nearly fifty years earlier; his initial skepticism about investing had long since evaporated. His holdings were earning 20 percent in the 1980s, enough to allow him to close his chain of New York furniture stores and retire with his wife to a beautiful house in Boca Raton, where they lived comfortably for years.

Richards was meticulous with his finances; he kept every statement Madoff ever sent him and every canceled check he'd ever written Madoff. Every time he received his quarterly statement, he checked the transactions against the stock prices in the papers, and they'd reliably match up. He had no reason to suspect that a man he had known personally for fifty years and who had been sending him checks for decades was running a massive con job.

So excited were the Richardses with the good fortune Madoff brought that they started looking for even more money to plow into their account. They decided to take out a second mortgage on their new home and invest the cash with Madoff. That put virtually every dime of their net worth in Madoff's hands. "I always said, 'Never put all your stuff in one basket,'" said Richards, "but when you have ten, fifteen, twenty, twenty-five years of experience with Madoff making money like clockwork, you forget about all those rules."

It wasn't just smaller investors who were caught up in Madoff euphoria. Thirty miles north, in Palm Beach, wealthy socialites celebrated as their millions grew into the tens of millions. Multimillionaires like Carl Shapiro were becoming billionaires—on paper—thanks to Madoff. In New York, the managers of feeder funds—Ezra Merkin, Thierry de la Villehuchet, and others—were making a fortune for their clients and themselves. If any of them harbored doubts about Madoff's legitimacy they never expressed them.

The virus eventually spread to the West Coast.

Gerald Breslauer and Stanley Chais were a different kind of Hollywood celebrity. They weren't glamorous—they were elderly men in their seventies in 2000—and they couldn't act, sing, or even direct. Yet they were Tinseltown royalty.

There are few jobs in Hollywood as valued and as embedded in the culture as that of personal business manager. Movie stars pay them to serve an almost parental role; they handle their bills, tell them what they can and can't afford, advise them on how to save their money, and tell them where to invest it. Directors, writers, and producers of a certain stature use them as well.

Everyone in Hollywood has a story about a business manager who took off with his clients' money, which is why Breslauer and Chais were so successful. The two competitors were among the town's trusted financial wisemen, with golden reputations.

Breslauer's clients were American icons like Barbra Streisand, Michael Jackson, and Bruce Springsteen and entertainment industry moguls like David Geffen and Barry Diller. As he grew closer to retirement, he whittled down his practice to focus on just two prized clients: Steven Spielberg and his partner at DreamWorks, Jeffery Katzenberg.

Chais represented his own elite roster of clients from his office in Beverly Hills. Among them was screenwriter Eric Roth, who won an Academy Award in 1994 for *Forrest Gump* and also wrote a number of Hollywood blockbusters, including *The Horse Whisperer, Ali, Munich,* and *The Curious Case of Benjamin Button.*

Both Chais and Breslauer earned large fees to invest their clients' money. And both, it turned out, were simply handing over the money to Bernie Madoff. Many clients claim they had no idea their business managers were letting Madoff do their work for them. Katzenberg, who entrusted millions to Breslauer, later said that he had never even heard of Bernie Madoff.

Stanley Chais's relationship with Madoff proved stratospherically profitable. His clients were making returns of between 20 and 24 percent annually in the three decades Chais and Madoff did business. But that was nothing compared to the returns Chais himself allegedly received. There were years when he reportedly made returns of up to 300 percent on his own investment. He and his various associates and companies allegedly withdrew a billion dollars from his Madoff fund between 1995 and 2009.

The fact that Chais's returns were so out of whack, and so much higher than his customers', should have set off clanging alarms for the money manager that something was terribly wrong. But Chais remained with Madoff until the bitter end, and continued to invest his clients' money as well.

These men were not the only Hollywood conduits to Madoff. Some of the most famous names in entertainment fell prey to Madoff's spell, including Kevin Bacon and his wife, Kyra Sedgwick, John Malkovich, and Larry King.

So much money, so few questions asked. The Madoff syndrome—joy over the profits he produced and a paralyzing inability to ask questions about them—was virtually universal among his investors. It was self-deception on a mass scale. Any one of his thousands of clients could have waved a red flag in front of regulators and requested an inquiry. Few of them did.

People see what they want to see when money rolls in. But Madoff's investors felt there was little risk. He had served as the nonexecutive chairman of NASDAQ. He was a Wall Street mandarin. To question him would be to question Wall Street itself.

Yet some people were asking questions about Bernie Madoff.

Michael Ocrant was the New York-based editor of a trade publication with a name almost designed to keep it buried in obscurity. *MARHedge* is short for "Managed Account Reports/Hedge Funds." He was a burly 42-year-old from Chicago, the son of a life insurance broker, without the kind of bravado or knack for self-promotion that might have landed him a reporting gig at a better-known publication. His readers were the managers, employees, and investors engaged in the hedge fund business, a once tiny niche of Wall Street that was now the most lucrative game in the market. Ignited by the phenomenal success of billionaire hedge fund investor George Soros, the industry was exploding. Worldwide assets shot up twelve-fold in ten years, from $39 billion in 1990 to $475 billion in 2000. Wealthy investors flocked to these private funds, which were free of government oversight and any constraints to invest conservatively. They were making millionaires and billionaires out of investors by plunging into areas where no mutual fund would dare tread: arbitrage, derivatives, program trading, short selling—the highest priced casinos in town.

Ocrant was on the outside looking in, working out of a dingy Manhattan office and armed only with a database jammed with facts about the industry. He was a numbers guy, skilled at mining hedge fund statistics for interesting stories. He had learned in college that every story is linked in some way to money, even on the crime and politics beats. "If you follow the money, it leads you to all sorts of things," he believed. The more the hedge fund business grew, the more intensively he ran the numbers.

In the winter of 2000, the reporter got the tip of a lifetime on a trip to a conference in Barcelona. Fate brought him together with Frank Casey, Harry Markopolos's colleague, inside a taxi from the airport. They were crammed so closely together they were almost forced to introduce themselves. Ocrant told Casey about his job. "I've uncovered a couple of the big scams around," Ocrant told him. "I know everyone in the hedge fund industry."

Casey's ears perked up. "Well, that's good, Michael," he said. "I'll make you a dinner bet that you don't know the biggest hedge fund in the world."

"Impossible," Ocrant said.

"Bernie Madoff Securities."

Ocrant was puzzled. "I've heard of Madoff Securities," he said, "but they don't run a hedge fund."

Casey explained that he was only technically correct. Madoff's investment fund wasn't registered with the government as a hedge fund, just as a broker-dealer. But his investment advisory operation was a private fund that was supposedly investing the money of wealthy people in complex, options-based trading, which is an almost textbook definition of a hedge fund. Because he had chosen not to register his investment advisory as a separate entity from his market-making operation, people like Ocrant had no idea he was even in the investment business.

Having lost the bet, Ocrant bought Casey and his wife dinner. But Casey had won something a lot better than a good meal: someone was finally listening to what he and Markopolos were saying about Bernie Madoff.

The editor realized he had a story regardless of whether Madoff's operation was legit. Madoff was a famous name on Wall Street, but primarily for his broker-dealer operation. The finance world would be amazed to learn that this old-school market maker was running a secretive hedge fund.

When he returned to New York, Ocrant went to work. He got hold of some of Madoff's investor statements from Casey and started to crunch the numbers. What they showed was astonishing: Bernie Madoff was running one of the world's largest hedge funds—perhaps the largest. Ocrant was amazed. How could people not have heard of this?

Phone calls started to fly between Ocrant and Casey. Why

were Madoff's returns so consistent, and for so long? Where was the volatility? Why was his split-strike conversion strategy so much more profitable than everyone else's?

The editor opened his address book and started calling around to contacts in the investment business. The consensus was that "no asset manager, not even one who was a computer programming genius given access to all the supercomputers in the world, could create a system that would produce such steady returns," Ocrant recounted. Many said they had smelled a rat and had outright refused to invest with Madoff. He called experts in options. Many told him they'd never found any evidence that Madoff was actually buying the options that were at the heart of the split-strike formula. The majority suspected that Madoff was running a scam.

Ocrant was all but reliving Harry Markopolos's experience from a year earlier, and he was coming up with the same findings. It had the makings of a huge story.

In April 2001, he called Bernie Madoff's office. True to form, Bernie returned the call.

Ocrant recounted in a *Times* of London essay that Madoff was surprisingly unthreatened by his inquiry. "Sure, I understand," Madoff said. "Why don't you come down now. I'll wait for you."

"Do you want to give me some days and times when you'll be available?" Ocrant asked.

"No, I mean why don't you come down now?" Bernie Madoff's charm offensive had begun.

A few hours later, Ocrant traveled up to the Lipstick Building and was met by the gray-haired eminence. The reporter was ushered through the nineteenth-floor trading area and into the Tank. For two hours Madoff put on a flawless performance. He was "courteous, composed, attentive and responsive," Ocrant says. He calmly declined to disclose the formula he was using to attain his superlative results; it was proprietary information, a "black box" system he'd invented into which the world wasn't allowed to peek, he said. "I'm not interested in educating the world on our strategy."

Competitors couldn't replicate his success because they lacked the physical plant and a large operation to achieve it, Madoff said in his street-smart New York accent. The massive order flow on his trading floor gave him "market intelligence" that informed his investment decisions. Critics looking for another answer were "wasting their time."

Madoff was so matter-of-fact in his explanation he was almost blasé. He was Bernie the teacher, patiently educating a young student.

At one point, Madoff waved in one of his sons to meet Ocrant. (Ocrant can't remember if it was Mark or Andy.) When the interview was over, Madoff proudly gave the reporter a tour of his trading floor and ushered him out the door.

Madoff had told Ocrant one lie after another. There was no black box with a secret investing strategy. The technology on his trading floor had nothing to do with the growth of his fund. He wasn't using market intelligence for his investment decisions. He was just stealing money and paying off clients so they wouldn't catch on. As he ticked off these fabrications, Madoff betrayed no nervousness and no remorse, just an easy, disarming charm. It was the behavior of a sociopath.

Yet his deceit was effective. He couldn't kill Ocrant's interest in the story, but he succeeded in planting a seed of doubt in the reporter's mind. "I kept thinking, God, maybe he really is an investment genius," he said. But Ocrant kept digging, and eventually he built a strong case. *MARHedge* published his article on May 1.

Madoff Tops Charts: Skeptics Ask How

Mention Bernard L. Madoff Investment Securities to anyone working on Wall Street at any time over the last 40 years and you're likely to get a look of immediate recognition.

But it's a safe bet that relatively few Wall Street professionals are aware that Madoff Securities could be categorized as perhaps the best risk-adjusted hedge fund portfolio manager for the last dozen years.

Most of those who are aware of Madoff's status in the hedge fund world are baffled by the way the firm has obtained such consistent, nonvolatile returns month after month and year after year.

[Hedge fund professionals] noted that others who use or have used the strategy . . . have had nowhere near the same degree of success.

The day after the article was published, Ocrant said, he received a phone call from a reporter at *Barron's* named Erin Arvedlund requesting a copy of the piece. A week later, she published her own, similar story in the widely read business weekly raising questions about Madoff and citing the *MAR-Hedge* article.

Arvedlund's piece, "Don't Ask, Don't Tell," quoted anonymous investors skeptical of Madoff's claims and suggested that his success was attributable to front-running, a form of insider trading. "Anybody who's a seasoned hedge fund investor knows the split-strike conversion is not the whole story," stated one. "To take it at face value is a bit naïve." Another investor said he had pulled his money out of Madoff when he grew suspicious of him. "What Madoff told us was, 'If you invest with me, you must never tell anyone that you're invested with me. It's no one's business what goes on here,'" the investor recounted. "When he couldn't explain how they were up or down in a particular month, I pulled the money out." A "very satisfied" Madoff investor was also quoted: "Even knowledgeable people can't really tell you what he's doing."

Arvedlund recounted a short phone interview with Madoff, who was on a boat in Switzerland at the time. He had clearly revved up the charm offensive once more. "He wasn't angry or upset," she recalled. "He sounded more than friendly."

"I can't really go into details," Madoff told her, his voice crackling over the long-distance connection. But he brushed aside the front-running rumor as "ridiculous." He scoffed when told his competitors couldn't replicate his results. "Whoever tried to reverse-engineer, he didn't do a good job," he said with contempt. "If he did, these numbers would not be unusual."

The cumulative effect of the *MARHedge* and *Barron's* stories should have been devastating to Madoff. The gist of both was that he was running a large, secret, and potentially illegal operation.

Back in Boston, Harry Markopolos celebrated. It seemed his year-long crusade had succeeded, and that he'd been vindicated. Bernie Madoff had been exposed. "Neil Chelo [a colleague], Frank Casey, and I felt one hundred percent certain that the SEC would be shutting down Bernie Madoff within days," he said. They waited for the inevitable calls to start pouring in from the media and SEC investigators.

But there was only silence. No one who could have done something to stop Madoff seemed to have read the stories, or cared enough about them to react. The SEC offices, it turned out, didn't even subscribe to *MARHedge* or *Barron's*.

The Madoff family, however, read them with great interest.

When *Barron's* hit the stands it created a buzz inside the Lipstick Building. Madoff's investment operation on seventeen was a source of mystery to the staff on the eighteenth and nineteenth floors, and employees watched Frank DiPascali with more than a little suspicion. Now the press was hinting that there was something illicit going on down there.

Mark decided to address the issue. He and Andy spent almost all their time overseeing the frenetic action on the trading floor and almost never talked with their employees about the investment advisory operation. But that didn't mean they were oblivious to it. The Madoff family spent virtually every waking moment talking business with one another, and the investment advisory operation was a huge source of revenue for the company. Now ominous stories were popping up in the press, and the brothers were undoubtedly asking their father questions of their own. But whatever they knew, Mark and Andy denied it all.

Mark stood up on the elevated dais on which he and Andy worked. "Can I have your attention please?" hollered the young, handsome 37-year-old. The floor hushed. "There's an article in *Barron's* that you may have seen," he said. "I want you to know that the allegations in it aren't true." There was nothing funny going on in the investment advisory division, he said. These were just the rantings of jealous competitors.

The traders were confused. "We always wondered how he did it," one trader said of Madoff's investment record. "But it was a separate business down there, and there weren't any clues. What am I going to do: outguess a guy with an impeccable reputation—a billionaire?"

As the family must have realized, the articles planted a seed of doubt in the minds of their employees. But their jitters likely subsided as time went on. Ocrant and Arvedlund never followed up on their stories. The SEC never called. Nothing happened.

● ● ●

When Harry Markopolos was in tenth grade in his Erie, Pennsylvania, Catholic school, he decided to pull a prank. His biology class was experimenting with fruit flies, and he pocketed a vial of flying insects and took it home. Alone in his room in his family's small, three-bedroom house, he spent the following days and nights breeding them, until there were tens of thousands of buzzing insects multiplying inside a five-gallon jar. It was a sight that might have frightened off other kids, but Harry wasn't your average kid.

He convinced his mother that he was conducting a special science project and had to be dropped off at school at 6:30 a.m. When he arrived, he walked into the building and headed for the empty cafeteria, where he proceeded to open the jar and release the massive swarm of insects into the room. Within three days, the entire school was infested. He never got caught.

Markopolos brought an oddball quality to situations. The combination of eccentricity and smoldering intelligence made him formidable, but always a little puzzling. His crusade to expose Bernie Madoff was based on his belief that Madoff was "a clear and present danger to our nation's capital market system and the reputation of the nation." He believed Madoff was "a domestic enemy." Army intelligence had left an indelible impression on him. But his self-characterization as fighting a kind of war of civilizations suggested not so much his military history as a Tom Clancy novel. His analytical skills had led him to make an extraordinary discovery about a competitor in the business world, but it was apparent that it also unleashed a wellspring of interesting personality traits.

"Unless you did the human intelligence-based fact gathering, you would not be able to see beneath the smoke and the mirrors, behind the curtains of secrecy," he explained. "I had a brave, highly-trained team that greatly assisted me. . . . I knew how to develop information networks and use those networks to identify key pieces of intelligence to go after, and I knew how to recruit people to a team to go after that information."

The images were straight out of Hollywood spy movies. In reality, his team consisted of four men with day jobs who were trying to expose Madoff in their spare time. Ocrant, who was running a magazine in another city, said the four of them never met together, but they all kept in touch about the case and helped one other in their efforts.

But Markopolos's penchant for melodrama went beyond his florid language.

From the start, his strongest ally inside the SEC was a twenty-five-year veteran of the agency named Ed Manion. The two had served on the Ethics Committee of the Boston Security Analysts Society and grown friendly. It was Manion who brought Markopolos in to the Boston office to make his case in May 2000, which had ended in a roomful of blank stares.

Soon after the terrorist attacks on September 11, Markopolos says, Manion called him to say he was concerned that the Madoff case had slipped through the cracks and needed to be pressed again. At Manion's urging, Markopolos resubmitted to the SEC his eight-page report from the previous year, along with some updated material. "These numbers really are too good to be true," he wrote the agency. "And every time I've thought a company's or a manager's numbers were 'too good to be true' there has been fraud involved." He added a section titled "What I Can Do to Help the SEC." One can only imagine what the agency's staffers thought when they read it.

"I can provide you with detailed questions for your audit team," he stated.

> In fact, I would be willing to accompany a team undercover under certain conditions (new identity, disguise, proper compensation) and willing to sign a nondisclosure agreement and serve under the command and control of the SEC.
>
> In return, I need complete anonymity. I would take a leave of absence from my firm. Only my wife would know where I am, but I would have no contact with her or anyone else that I know while on assignment. I feel that my personal safety of myself and my family may be in danger if I assist the SEC.

Markopolos's cloak-and-dagger mentality jumped off the page. It was hard to focus on his otherwise brilliant deductions about Madoff's operation. Once again, the SEC never responded to him or his accusations.

Markopolos did not interpret the inaction as a mere rebuke. He worried that Madoff had been tipped off by friends in the SEC and was out to kill him. "I know that, if you're a securities lawyer, and someone hands you a bunch of derivatives math proofs—I submitted them to the SEC in that submission—they're going to have plenty of questions because they're not trained in that kind of math," he recalled. "And I received no questions. So I panicked. I feared for my life." He said that he went to a police station and developed "security protocols" with them. "They'll tell you how frightened I was," he recalled. "It was a terrifying experience."

At this point, he estimated, Bernie Madoff's Ponzi scheme encompassed over $10 billion in investors' money. Markopolos was outraged that the SEC was dropping the ball, even though he spoon-fed them the facts. But the messenger was complicating his message.

It was a shame, because no one understood the dynamics of Madoff's scam as well as Markopolos did. Years later, he explained why so many of Madoff's investors refused to question Madoff's methods:

> Madoff was not transparent. He would never let you behind the magic curtain, where the black box was. He would just tell you he had a black box, he couldn't let you look inside the black box, because what was taking place inside the box was so secret that once you saw it and understood it you'd be able to duplicate it.

And that was his cover story. And a lot of people never wanted to open that box because they wanted those returns. They wanted to believe that those returns existed, but of course they did not. And they cast aside all suspicions and all doubt and all fear, and they let greed overrule all else.

And that's why it ended in disaster, unfortunately.

8

The World at His Feet

On paper, Sherry Shameer Cohen seemed a logical hire.

In January 1987, Walter Noel, the patrician owner of a small Greenwich, Connecticut investment firm, lost his assistant to another company and was casting about for a replacement. On the recommendation of his departing secretary, he gave the job to Cohen, a former colleague of hers.

If anything, Cohen was overqualified. She had held an administrative job at Paine Webber for six years and had a Series Seven stockbroker's license, which allowed her to fill stock trade orders when clients called the office. But temperamentally, it was an almost comical mismatch.

Noel was the silver-haired patriarch of a large Greenwich family, the father of five perfect-looking daughters, most of them blonde and all of them model-gorgeous. *Vanity Fair* published a profile of them in October 2002, apparently because they embodied the WASP ideal. "The five Noel women have made a name for themselves by shoring up the virtues of a nearly extinct aristocracy," it gushed. "They're well-educated and well-married, and they're raising a pack of well-behaved, multi-lingual children while keeping their string-bikini figures intact." The daughters were captured in a Bruce Weber photo, laughing arm-in-arm in a field of grass, as if recreating a shampoo commercial.

Cohen, by contrast, was an overweight, middle-aged Jewish woman from the Bronx, outspoken, opinionated, and uncomfortable around bluebloods. She was the ugly duckling in a pond of beautiful swans, never to be accepted, confided in, or solicited for her opinions. "The family treated me like the help," she said.

She was good at her job, though, and Walter developed a soft spot in his heart for her, keeping her employed for more than a decade. As a result, she witnessed the beginnings of one of the great financial train wrecks in history. She observed with amazement as the Noel family ran into the arms of Bernie Madoff, who set them off on a delirious, cash-fueled romp across the globe, clueless about the workings of his wealth machine but high on the money it produced and their ability to flaunt it. The Noel family's Fairfield Sentry Fund became the single largest feeder to Madoff's investment operation, even though Fairfield officials were almost oblivious to how it operated.

When Cohen went to work for Fairfield Greenwich Group, it was a tiny firm, consisting of Walter, his daughter

Corina, and a few other employees. The Noels were a family with high social aspirations, and Walter and his wife, Monica, were living a life they could barely afford, borrowing against their house to pay for their country club memberships, Rolex watches, and vacations in Brazil and Switzerland. "The unofficial motto was, 'If we act rich, we will become rich,' " Cohen said.

Walter was a dignified and courteous southern gentleman, raised in Nashville and educated at its finest educational institutions, Montgomery Bell Academy and Vanderbilt University, before heading off to Harvard Law. Soon after graduating, he met Monica, a Wellesley girl who was the product of a wealthy Brazilian family. The couple raised a family in cities around the world as Walter pursued private banking positions in Nigeria, Boston and Switzerland. He opened his boutique investment firm in Connecticut in 1983.

By the time Cohen came to work for him, Walter was 57 years old and still mowing his own lawn and buying office furniture secondhand. He was investing like a proper conservative businessman, making safe choices and eschewing the risk taking preferred by his fast-lane competitors. Every Friday, he'd get a copy of the tip sheet *Value Line*, pour through its weekly stock picks, and choose his investments from among them. It made for very predictable days at the office.

But his desire for wealth never faded. Some people have to wait decades before a big opportunity presents itself. Walter's chance arrived in 1989, just steps away from his office door.

Another investment company, Fred Kolber and Co., sublet some space from him, and Noel watched its investment performance over the years with admiration. He was particularly impressed with Jeffrey Tucker, one of the owners, who was a former SEC lawyer. The Tennessee WASP and the Jew from Queens struck up a friendship and eventually joined forces.

Tucker wasn't part of Noel's world; he didn't dress nearly as well and didn't have the same world traveler's sophistication. Cohen thought Tucker was a bit envious of his new partner's lifestyle. "I always thought that he kind of had his nose pressed to the glass," she said. But their alliance changed their lives. Tucker's father-in-law knew Bernie Madoff, and Tucker suggested to Noel in 1989 that they meet with the famed investor. "I've got this guy who's got really impressive returns," he told Noel. "Let's look into it. Maybe there's a product we can develop."

Like so many others before them, the two fell for Madoff in a big way. They were dazzled by his stature, floored by the astonishing consistency of his investment returns and seduced into believing in the alchemy of his split-strike conversion formula. Tucker had spent nine years at the SEC and was more qualified than most to do proper due diligence. But the enormous secrecy shrouding Madoff's fund, the lack of independent audits, and its surreal, nearly perfect performance year after year raised no red flags with him. The evidence that jumped off the page at Harry Markopolos—the impossible number of options required to carry off Madoff's strategy, for example—either escaped the former regulator's notice or meant nothing to him. Despite Tucker's fatal oversights, the company thereafter heralded his background at the SEC as proof of their rigorous scrutiny of Madoff.

The budding alliance led to the debut of the Fairfield Sentry Fund in 1990, which was invested exclusively with Bernie Madoff. The price of admission for the fund's investors was 20 percent of the profits Madoff produced (and later 1 percent of the assets placed in the fund). It was no surprise that when Walter Noel and Jeffrey Tucker advertised Madoff's performance record, investors started flocking to them.

If Madoff was a boon to this small Connecticut firm, Noel and his family were a godsend to Bernie Madoff. A Ponzi scheme is a voracious animal with a never-ending appetite for cash. The longer investors remain in the fund, the more likely they are to draw their profits out of it. Madoff could keep his scam afloat only by continuously feeding the animal fresh money. Luring checks out of rich Jews on country club golf courses was no longer cutting it. Walter Noel's family was a one-family international social register; their feet were planted on several continents and their friends were members of the jet set. His potential to raise international investment dollars for Madoff was boundless.

His ticket to fortune was his beautiful daughters.

Vanity Fair was right: the Noel girls were "golden." Monica Noel had been raised in the rarified world of Brazilian aristocracy and handed her values down to her children. The five Noel daughters were cultivated in far-flung European cities, moving wherever Walter's globe-trotting jobs took them. They were taught to live their lives with a certain level of style, from the way they dressed to the manners they displayed when company came for dinner. "Mom would say to us, 'Join in and think of something interesting to say, keep your posture straight—but first go upstairs and brush your hair again,'" one sister told the magazine.

Their educational pedigree was flawless. Corina went to Yale, Lisina and Ariane went to Georgetown, Alix went to Brown, Marisa went to Harvard. The men they married were East Coast preppies and foreign-born members of the international jet set. All the men were from prominent families. Corina's husband, Andrés Piedrahíta, was the son of a wealthy Colombian commodities trader. Lisina married Yanko Della Schiava, whose mother was the editor of *Cosmopolitan* magazine in Italy and whose father was the editor and publisher of *Harper's Bazaar* in Italy and France. Alix married Philip Toub, son of a major Swiss shipping magnate.

They were a formidable family, wealthy, glamorous, and beautiful, with roots deep in the world of the super-rich.

Cohen never cared for Piedrahíta. "He was slick," Cohen said, "someone I didn't trust. He smiled a lot. He was polished—overly polished." But his arrival in the Noels' life was a big deal.

"I love people," he told the *Wall Street Journal* in 2009. "I am known for being fun." No one would argue with that. Corina's husband was a force of nature, a short, handsome businessman with a big, friendly smile and a natural charisma. He'd spent much of his life hanging out in monied European circles.

Piedrahíta's enormous self-confidence was a blessing and a curse; it won him a legion of friends and business contacts but led him to take huge risks. After studying communications at Boston University, he plunged into the world of finance in the early 1980s with little training but a host of rich friends to hit up for his investment projects. Working as an investment advisor for a small commodities dealer named Balfour Maclaine, he set out with characteristic zeal to sell his family's friends a few investments, which ended up going bust. He lost eight clients a total of about $600,000 and left the company.

But it was just a setback for a man who was born to cut a swath through the world. Piedrahíta bounced around a number of investment firms before opening one of his own, Littlestone, a loose translation of his last name, in midtown Manhattan. In 1989 he married Corina. Just after that, her father discovered Bernie Madoff. The burgeoning Noel family was beguiled by the mystique of the famed investment manager and his split-strike conversion strategy, so much so that all but one of Noel's sons-in-law went to work as global salesmen for Noel's Fairfield Sentry fund.

As she watched their fortunes rise, Sherry Cohen felt the Noels barely understood the product. "I don't think they ever really questioned the split-strike thing," she said, "as long as they saw returns coming in." When the paperwork from Madoff Investment Securities arrived on her desk, she noticed the same oddities that struck John Pohlad, the investment advisor in Minneapolis. Madoff's statements were shoddy: the paper stock was cheap, there was no splashy corporate logo on them, and there was none of the usual fine print with customer service telephone numbers to call for assistance. They were apparently spit out of a typewriter or a dot-matrix printer.

Cohen asked Walter if she could transfer the information onto a spreadsheet. What it showed was at once conventional and strange. "It was a very, very conservative portfolio, not unlike my own, except they had a lot more holdings," she said. "But I kept thinking: 'I don't understand this. Is it possible that I chose the wrong stock in every single industry so that my own returns are about half of what he's getting?'"

Cohen says she brought up her concerns with Corina, who ran the office with her father. Madoff's was a conservative fund, Corina explained, one that slightly increased in good times and barely fell in bad times. She thought little of Cohen's worries.

As it was, Cohen was little more than a glorified secretary, and no one much cared about her opinions. "If you weren't there to make money for them it wasn't worth their time," she said.

Noel's foreign-born sons-in-law went about carving up their sales territory around the globe. Piedrahíta took Latin America, Britain, and Spain; Toub took Brazil and the Middle East; Della Schiava took southern Europe. The money the family raised over the next decade dwarfed anything Madoff had seen in the 1970s and 1980s. The river of investment money streaming into the Lipstick Building's seventeenth floor turned into a floodtide. They successfully promoted Fairfield Sentry to wealthy individuals and to major Swiss banks, such as Union Bancaire Privée and Banque Bénédict Hentsch in Geneva. Although they didn't invest in Fairfield directly, banking giants HSBC, BNP Paribas, Fortis, and the Royal Bank of Scotland poured billions of dollars into loans to hedge funds looking to invest in Fairfield Sentry.

It wasn't just that the Noel family knew so many important people; the product they were peddling was also an easy sell. Bernie Madoff's reputation had spread overseas, and the royals and bankers of Europe were just as eager to get in on his elite fund as were the company presidents lounging in the dining room of the Palm Beach Country Club.

After sixty years, Walter Noel finally became a gloriously wealthy man. Fairfield's commissions and fees from Madoff ran north of half a billion dollars between 2003 and 2008. The money circulated through an offshore investment company named Citco—one of Fairfield Greenwich's offices was located in Bermuda—enabling the company's foreign investors to avoid paying U.S. taxes.

The Noels bought homes in the playgrounds of the superwealthy: Palm Beach, Park Avenue, Southampton, and the island of Mustique. "Easy in the islands," raved a cover story in *Town and Country* magazine in May 2005. "For Monica and Walter Noel, their hilltop retreat on Mustique is all about the mix—of family, friends, great times and a sexy global design style."

It was a grand time for the family. Andrés and Corina traveled on a Gulfstream jet and lived in a mansion in Chester Square in London before moving to Madrid. They also owned an estate on the Spanish resort island of Mallorca, with a Falcon yacht for their boating pleasure. The guests at their dinner parties were often members of European royal families, including the Duke of Marlborough and Spain's Prince Felipe. Many of the royals went on to become investors in Fairfield Sentry. "I'm a connected guy," Piedrahíta told the *Wall Street Journal* in 2009. "I know a lot of people all over the place. I have the ability to make friends."

He was more visible than the other Noel men but was hardly its only successful salesman. Philip Toub, Alix's husband, scored a mammoth coup by persuading the Abu Dhabi Investment Authority, the world's largest sovereign wealth fund, to invest an astonishing $400 million with Madoff via Fairfield Sentry in 2005. Bernie had come a long way from collecting retirement checks from retirees in the Catskills.

Fairfield Sentry ultimately raised billions for Madoff, making it the largest single contributor to his investment advisory business.

A firm collecting all that money for just one investment manager might have thought to check him out pretty thoroughly. Noel and Tucker assured Fairfield Greenwich investors of their thorough vetting techniques. "The nature of

FGG's manager transparency model employs a significantly higher level of due diligence work than that typically performed by most fund of funds and consulting firms," the company's website claimed. "This model requires a thorough understanding of a manager's business, staff, operational practices, and infrastructure. . . . FGG's business model enables the firm to have privileged access to all aspects of a manager's operation and investment process, including security level transparency for risk monitoring purposes."

Far from having a "thorough understanding" of Madoff's practices, though, Noel, Tucker, and their glamorous salespeople seemed hopelessly vague about how his strategy worked. Several investors walked away from meetings with them so concerned about the Fairfield executives' lack of understanding of the split-strike strategy that they refused to invest with them. When a London money manager named David Giampaolo attended a presentation Piedrahíta gave about Fairfield Sentry in 2007, Piedrahíta was unable to answer basic questions about the strategy. "There was no deep scientific or intellectual response" to his questions, Giampaolo later told the *Wall Street Journal*. A client of Giampaolo's was so alarmed that he decided not to invest.

Fairfield Greenwich also promised investors that it was regularly tracking Madoff's investments. But the trade tickets he sent Fairfield were three to five days late and didn't even have trade prices on them, just daily averages. According to a suit against the company by the Commonwealth of Massachusetts, his statements did include trade prices, but they were often wrong. In some cases the transactions would have fallen on weekends and holidays, when the markets were closed (the company disputes it).

As for the firm's "privileged access," the truth is that Noel and his firm had limited access to Madoff's operation. They'd been given a tour of his trading floor but had never seen the seventeenth floor, where their investments were allegedly being handled, or seen anyone there actually carrying out the split-strike strategy.

There was a litany of potential tip-offs to Madoff's fraud, from the SEC's crackdown on Avellino and Bienes in 1992 to the articles raising questions about Madoff in *Barron's* and *MARHedge.* In fact, Tucker's curiosity was piqued by the *Barron's* story, and he paid a visit to the Lipstick Building in the spring of 2001 to ask some questions. He was granted a meeting with Madoff and Frank DiPascali, who proceeded to snow him with ease. They showed Noel's partner some papers that Madoff claimed were records of all his trades. Madoff then gave him another journal, which he called the "stock record" for Madoff Securities. "Pick any two stocks," Madoff told him. He might as well have been performing a card trick.

Tucker picked AOL/Time Warner out of the booklet. Madoff sat down at his computer and pulled up the AOL listing from the website of the Depository Trust Company, which tracks the actual holdings of clients of U.S. securities firms. "He or Frank activated a screen that he said would get into their DTC account," Tucker said. "And they continued to move pages of the screen until they got to the AOL page and with the stock record I could compare the total number of AOL [*sic*] according to the Madoff stock record with the Madoff account at DTC which had the—which tied. . . . And I knew of my own that the position I saw for us was roughly what we had because I was somewhat familiar with our size, you know, in shares. . . . That was basically the meeting." Tucker was satisfied. "Everything checked out," a company official would tell the *New York Times* years later.

But Tucker had never seen a DTC screen before. And he apparently never verified the documents Madoff handed him or tried to independently confirm the trades Madoff had produced. If he had, he would have discovered that the trades were all fiction.

Yet another clue that something was amiss came in 2007, when Noel, Tucker, and Piedrahíta, who had become a partner in the firm, shopped the company around for a possible sale or public offering. The effort came to a screeching halt when Madoff refused to allow anyone to go through his books.

Noel wasn't alone in sticking his head in the sand. Ezra Merkin, Stanley Chais, and other feeder fund executives had much the same evidence sitting before their eyes. There was simply no incentive to kill the golden goose. Madoff had transformed the Noel family's life; they couldn't have done better if a gusher of oil had exploded on their front lawn. Life for them had become a continuous party since they'd met him.

As for Sherry Cohen, she was never invited to participate in the festivities. For eleven years, she made the coffee and answered the phones back at the office. She did all her work aware that in an office filled with globe-trotting millionaires, her role was to be invisible. "They worshipped money," she said.

It all ended for her in 1998. Fairfield Greenwich expanded to Manhattan as a result of its growing good fortune, and she was let go. Walter said the company had outgrown her abilities. Cohen didn't believe him. "They wanted a good image," she said. "They thought I was fat."

Bernie and Ruth Madoff loved Europe. It was a world that valued discrete wealth, and they were nothing if not discrete and wealthy. When they traveled there, they had routines and rituals, favorite hotels, and regular restaurants.

As in Palm Beach, the couple seemed to enjoy breathing the air of the privileged while attempting to avoid drawing attention to themselves. They owned an apartment in Cap d'Antibes in the South of France, a vacation getaway for the superrich where you could sometimes spot Madonna jogging around the cape accompanied by bodyguards. But their home was a modest, 1,300-square-foot apartment with an obstructed water view. Typically, they'd forsake socializing with their fellow millionaires and dine out quietly over a lobster salad or sea bass at Les Vieux Murs, a seaside restaurant in Old Antibes.

But their private ways hardly precluded their indulging

in the luxuries that Europe affords the monied class. In Paris they stayed at the Hotel Plaza Athénée near the Champs Elysées, a favored spot of the cultural elite for generations, from Ava Gardner to Yves Saint-Laurent. Bernie became so chummy with the concierge there that he booked the man's restaurant reservations when he visited New York.

The Madoffs flew from Paris to London in a private jet and relaxed in their usual suite at the Lanesborough, London's most expensive hotel. Bernie kept a separate wardrobe in the room, lest he arrive in London without a dress shirt with the necessary high-cotton weave. At night, he and Ruth liked to dine at The George, where Shakespeare and Dickens used to come for drinks, and The Ivy, a venerable celebrity hangout. If Madoff had any qualms about living so well on stolen money he never showed it.

As the years went on, Bernie's operatives and feeders multiplied beyond the world of Noel and his sons-in-law; his European business exploded in the 1990s and the first decade of the twenty-first century. Large banking institutions such as Spain's Banco Santander invested over $3 billion of its clients' money with him, and a new feeder fund managed by Bank Medici in Vienna raised a similar amount.

Bank Medici's founder, Sonja Kohn, was the picture of aristocracy. Hailed as "Austria's woman on Wall Street" at investment events, she was a member of the European financial elite. Like many of Bernie Madoff's impresarios, though, the mystique was something she'd largely invented. Kohn grew up in a middle-class Austrian household and moved to the

United States to raise five children with her husband, Erwin, settling in the Orthodox Jewish hamlet of Monsey, New York. In 1984, at the age of 36, she passed her first broker's exam.

Her moment came when she met Bernie Madoff, who bestowed his Midas touch upon her and sent her on her way to raise money for his investment fund. As it had for Walter Noel, Robert Jaffe, and a dozen others, the benediction allowed Kohn to transform into a far grander version of herself. In 1990, she founded Eurovaleur, an investment banking firm that marketed accounts with Madoff. Her offices were in the towering General Motors Building, overlooking Fifth Avenue and Central Park. In 1994, she returned to Vienna and founded Medici Finanz, renamed Bank Medici in 2003. Its headquarters was steps away from the Vienna State Opera and was decorated to evoke an Old World ambience, filled with dazzling Baroque furniture. Clients were captivated by the impression that Medici had an extraordinary legacy. It was a huge success, and Kohn soon became a high-profile advisor to Austria's minister of economic affairs. As a token of the country's appreciation, she was awarded the Grand Medal of Honor for Service to the Republic of Austria in 1999.

But Bank Medici operated in a bizarre world of its own. It had no connection to the storied Medici family, the house that produced the most powerful bankers of the Renaissance, three popes, two queens of France, and several cardinals of the Roman Catholic Church. Medici was nothing more than a "fantasy name," Kohn said in a 2004 interview with the Austrian newspaper *Wirtschaftsblatt*. She had branded her company brilliantly, masking an operation with major shortfalls in experience and investment savvy.

Kohn never missed an opportunity to promote Medici's flawless investment track record, but it was all thanks to Bernie Madoff; she had entrusted an astounding $2.1 billion of her clients' money with him and bolstered her company's prestige. She marketed the bank as the gateway to Madoff's prosperity, solidifying herself as one of the financial mastermind's largest conduits. She attracted wealthy Europeans, Israelis, Arabs, and even powerful Russian oligarchs to her funds, contributing a staggering amount of money to Madoff's widening net across the Atlantic. Her fortunes would rise and fall with his.

In 2000, Madoff expanded the size of his London office, located in a townhouse on a street in the Mayfair district known as "Hedge Fund Alley." He had opened Madoff Securities International Ltd. in 1983 as a small operation geared toward trading stocks overseas, but in 2000 he ramped up the operation, loaning it $62.5 million and adding traders to its operation, according to the *Wall Street Journal*.

Madoff's travels to Europe grew more frequent after 2000, and his strange compulsions traveled with him. The staff at his London office grew familiar with his various obsessions and tried endlessly to anticipate his concerns. "We'd spend days before his arrival leveling the blinds, making sure the computer screens were an identical height, lining every picture up straight," his office manager, Julia Fenwick, told London's *Daily Mail*. "No paper was allowed on the desks.

We'd use black marker pens to touch up the skirting boards and the doors. Anything that looked as if it had a mark or scratch on it, we'd have to retouch. Things like that would drive him nuts."

He mandated that the London office mirror the look of the New York office; everything had to be decorated in black and gray, from the carpets to the closets. He had a fetish about straight lines; he threw the staff into a tizzy after a brand-new video conferencing system was installed on a rounded wall. "I can't have that in my office," he told staffers. "It has to be square." The wall was demolished soon afterward.

His obsessions weren't limited to the office decor. Madoff had surveillance cameras installed inside the office to allow him to watch from his computer in New York, to see if the London staffers were taking long lunches.

Fenwick said that Madoff often had tailors from Kilgour on Savile Row come to the boardroom to take his measurements as Ruth sat at a desk "knitting bootees for one of her grandchildren." He was as fastidious about his $24 million Embraer Legacy Jet as he was about his clothing; Fenwick said he had it painted black and gray to match his offices and forbade anyone from bringing aboard luggage made of metal, lest it damage the interior. Peter Madoff enforced the rule when he traveled with his brother. "You can't put that there, you might mark something," he would tell passengers. "Bernie would kill me."

Madoff was forever trying to maintain order at his multibillion-dollar criminal enterprise, as if he could control the ticking time bomb he'd created by keeping the blinds straight.

The London office was a critical organ in the company's circulatory system. Back home Madoff was claiming that trades for his nonexistent split-strike conversion machinery were being executed in Europe, a story that seemed to be swallowed whole by virtually all his traders, managers, and family members in the Lipstick Building. To bolster his story, he regularly funneled millions of dollars from his investment advisory operation to his office in London and then bounced the money right back to New York, a round-trip transatlantic money-laundering exercise.

The traders at the London office spent their days playing the European stock markets for the company's legitimate proprietary trading side and had little contact with their New York counterparts. Yet it's far from clear whether they asked questions about a split-strike operation they were supposedly executing.

The money-laundering operation Madoff was running through London was a precarious operation with a hundred potential pitfalls for him. It was enough to drive a nervous swindler a little batty. One can only wonder how he would have reacted if he knew someone was on his tail.

The Noels had been peddling their Madoff feeder fund in Europe for twelve years when Harry Markopolos took his own trip to Europe in June 2002. He was there on company business, selling an options-based arbitrage product that bore a resemblance to Madoff's split-strike conversion strategy. Markopolos's algorithmic formula produced bigger profits but also a higher risk of failure than Madoff's. And his product was legit, Markopolos thought to himself.

He was traveling with Thierry de la Villehuchet, the chief executive officer of Access International who had sparked Markopolos's crusade by bragging to Frank Casey about his Madoff connection. By now de la Villehuchet had 45 percent of his funds invested with Madoff, over a billion dollars from the many European counts and countesses, dukes and duchesses he advised. His clients included Liliane Bettencourt, the daughter of the founder of the L'Oreal cosmetics empire and one of the world's richest women, as well as members of the Rothschild family.

Markopolos's Boston-based company, Rampart Investment Management, was a competitor of Madoff's, but Access International was free to partner with other companies, and de la Villehuchet agreed to market Markopolos's product to companies in Europe. He took Markopolos on an exhausting ten-day, four-city tour and held more than a dozen pitch meetings with private client bankers and hedge fund advisors. Each time, Markopolos sat with de la Villehuchet or one of his Access International colleagues: Philippe Junot, the former husband of Princess Caroline of Monaco; Prince Michael of Yugoslavia, the French-born grandson of King Umberto of Italy; and Prince Paul of Yugoslavia. Their one common bond, Markopolos later wrote, "was that their ancestors served as Napoleon's field marshals."

The meetings all started the same way. De la Villehuchet or one of his partners would introduce Markopolos and describe his product in a way they felt the bankers could understand. "It's just like Bernie Madoff, only it's higher-return and higher-risk," one of them would say, arguing that Markopolos's product represented an alternative way to employ an options strategy and diversify their holdings.

It made Markopolos furious: "Every time they said this, it was all I could do not to jump up and say, 'hey you morons, Bernie's a Ponzi so I have higher returns than his and lower risk because my returns are real while Bernie's are a fraud.'"

Instead, he sat quietly steaming as the bankers around the table expressed their infatuation with Madoff. "We have a special relationship with Mr. Madoff," someone would invariably say. "He's closed to new investors and he takes money only from us. . . . We're so proud of that relationship, and it's so meaningful to us."

Markopolos was an extra competing against a star. The bankers were elated that they had gained special entrance into Madoff's elite fund. It was the kind of giddy reaction that novice investors like Donny Rosenzweig had when Madoff opened the doors for them. But these were major European bankers who should have known better. Markopolos was dumbfounded. "They're going to get wiped out," he thought.

But he didn't feel comfortable tipping his hand and telling these bankers that Madoff was a crook. He worried that many European clients were investing with Madoff via offshore feeder funds—"only one step removed from organized crime," he called them—and Madoff's life would be in jeopardy from his mobster clients if his scam were exposed. Madoff would rather kill him than risk being rubbed out, Markopolos theorized, his cloak-and-dagger logic running in overdrive. "If I had jumped up and told everyone I met in Europe that Madoff is a fraud, I would not have gotten past the first meeting because the Access people would have dumped me on the nearest curb and then told Madoff that I was dissing him before his investors," Markopolos stated. "Madoff would then likely have had me killed—and he had literally 20 billion reasons why . . . As it was, I felt I was living on borrowed time."

Not every bank in Europe was caught up in the frenzy to invest with Madoff. Two years earlier, when Markopolos first learned about Madoff, executives of Credit Suisse Group AG held a meeting in New York with Madoff and some Fairfield Greenwich executives. The bankers flagged many of the problems that others had not, including the lack of an independent custodian for Madoff's assets and the poor auditing mechanisms at the company. The meeting only made Credit Suisse more alarmed because Madoff characteristically refused to give the bankers details about his investment operation. They left so befuddled about his methodology that they urged the bank's clients to pull their money out of his fund. Customers withdrew $250 million as a result.

A similar scenario played out in 2003. Bankers from Société Générale in France traveled to the Lipstick Building to perform due diligence on Madoff. Once there, Madoff's employees described the split-strike conversion strategy for them. The problem came when the bankers returned to their offices and repeatedly failed to recreate Madoff's results. They immediately placed him on the company's investment blacklist.

Access International felt no such reservations about Madoff. The company waived many of its vetting rules out of respect for his estimable reputation on Wall Street. "Of course we made an exception with Mr. Madoff," said Patrick Littaye, a partner of de la Villehuchet's. "I can't imagine asking him to pass a handwriting test."

Markopolos believed that de la Villehuchet was in over his head. "I found him to be a wonderful salesman without any discernable quantitative finance skills," he said. He claimed that he tried to warn de la Villehuchet about Madoff but was too worried for his own life to say too much to him. "You know that Madoff's a fraud don't you?" Markopolos said he told him. "Those returns can't exist." But de la Villehuchet wouldn't budge.

"I wasn't going to reveal that I had turned [Madoff] in to the government, twice already to the SEC," Markopolos said. "I was never going to say that because that was like signing my own death warrant. It would have killed me if I had come forward and been too open about it." Yet even if he had told de la Villehuchet about his submissions to the SEC, it is not certain that the Frenchman would have cared.

In the years that followed, de la Villehuchet's enthusiasm

for Madoff only grew. He increased his holdings with Madoff dramatically, ultimately investing 75 percent of his clients' money, and his entire fortune, with Bernie.

Markopolos never received as much as one death threat in his long crusade to expose the Ponzi scheme. It was de la Villehuchet who would pay the price for Bernie Madoff's betrayal.

9

See No Evil

On a cold December night in 2002, New York State Attorney General Eliot Spitzer took a breather from his crusade against corruption on Wall Street and traveled to Boston.

He had accused some of New York's biggest investment houses of peddling misleading stock research to benefit their clients, and rumors were flying that he was on the brink of a settlement. But this evening Spitzer took a respite from his relentless campaign and agreed to appear at the John F. Kennedy Presidential Library for a lofty conversation about corporate responsibility.

Wearing his banker blue suit and his famously purposeful demeanor, he settled into his seat alongside the CEO of Starbucks and the former president of Nike. The moderator, Harvard Business School professor Rosabeth Moss Kanter, geared most of the questions to the man who had been making life miserable for the country's financial institutions. Grinning mischievously as she goaded him to debate, she asked whether American corporations were being scapegoated for the country's problems.

Spitzer scoffed at the idea. "We are not scapegoating business for failing to resuscitate our educational system," he replied. "We're scapegoating business for building the Ponzi scheme instead of a decent business model."

Quietly slipping into the small crowd of scholars, reporters, and members of the public was Harry Markopolos. He was brandishing an envelope containing evidence that Bernie Madoff was running the biggest Ponzi scheme in history. Markopolos knew all about Spitzer's reputation as the scourge of corporate miscreants and had decided to seek his help. The documents in his envelope were the same ones he'd given the puzzled officials at the SEC's Boston office two and a half years earlier.

But Markopolos was frightened. He had heard that Spitzer's wealthy family was invested in hedge funds, and he guessed that the attorney general was a Madoff investor himself. (Spitzer's family lost an undisclosed sum.) Markopolos worried that Spitzer might be in league with Madoff—another reason to fear for his life. So he handled his envelope with white gloves to avoid leaving fingerprints. Instead of handing it to Spitzer himself, he gave it to a museum staffer. "I handled it only with gloves because I thought that he was an investor, and it turned out the *New York Times* said that he was," Markopolos said later.

He fled the library without waiting to see if Spitzer got the envelope. Spitzer later said he never did. "I'd remember meeting a guy wearing white gloves," he said.

The episode was a perfect illustration of why Bernie Madoff was able to run his criminal enterprise for so many years without getting caught: his only nemesis was an eccentric mathematician with a paranoid streak, and the government agency responsible for checking out Markopolos's claims was ignoring him.

The Securities and Exchange Commission was weathering hard times at the start of the twenty-first century. Spitzer's relentless crusade against securities fraud exposed the agency as a passive and somewhat feckless regulator. Critics were complaining that the Bush administration's pro-business policies were emasculating the once revered industry watchdog.

While Markopolos was virtually banging at its doors, the agency occasionally poked around to see what was going on at Madoff headquarters. In early 2004, nearly three years after the *MARHedge* and *Barron's* stories were published, a lawyer from the SEC's office in Washington was assigned to investigate Madoff's dealings with hedge funds, according to the *Washington Post*. Genevievette Walker-Lightfoot was a staff lawyer who had joined the agency well-armed; she was an expert in trading strategies from her previous job at the American Stock Exchange. After reviewing how Madoff executed trades for his investment advisory operation, she couldn't make sense of how his strategy worked. The contradictions in the information Madoff sent to her office were endless.

In March 2004, she went to her supervisors with her concerns. According to the *Post*, she went to Mark Donohue, one of her department's branch chiefs, and Donohue's boss, Eric Swanson, an assistant director of the department. As it turned out, Swanson had just met Bernie Madoff's niece, Shana, at an industry breakfast in October 2003. The two would see each other occasionally at industry conferences while Swanson was still employed by the regulator, although Swanson and Shana claim that their romantic relationship didn't begin until April 2006, shortly before he left the SEC. After Swanson left the SEC in 2006, he proposed to Shana on her birthday, December 8 of that year.

After Swanson and Shana were married in 2007, Bernie bragged at a business roundtable about his close relationship with SEC regulators. "There's always this friction between the regulation side of the industry and the practitioners about where you draw the line," he said. "I'm very close with the regulators. . . . As a matter of fact, my niece just married one."

One month after Walker-Lightfoot raised suspicions during the 2004 investigation, Donohue instructed her to put aside the Madoff case and delve into a probe involving mutual funds, the *Post* reported. A few weeks later, Donohue asked Walker-Lightfoot to turn over all her Madoff materials to a coworker, who boxed up the case and sent it to New York. The New York office looked into Walker-Lightfoot's investigation but only slapped Madoff with three minor violations in 2005. No evidence of fraud was found, and Walker-

Lightfoot's expertise was never again sought in the Madoff case.

Bernie became alarmed each time the SEC came to visit his office and insisted on dealing with its investigators personally. He ensured that when they came calling, his assistants received them on the eighteenth floor and escorted them upstairs to his nineteenth-floor office. They never saw the investment operation on seventeen.

But he needn't have worried. He and his aides were amazed at the youth and inexperience of the regulators who showed up at their door. "They would walk in and we'd look at them and we'd say to each other, 'What do you think their combined age is, twelve?' Elaine Solomon said. "They'd send kids. I think it was their first job out of school." Madoff found them easy to charm. "They were so impressed with the surroundings," Solomon said. "They were impressed with Bernie, the former head of NASDAQ, this wonderful man. Peter and Bernie would go in and chat with them, you know, 'You can go here for lunch or you can go there for lunch.' "

In response to their queries about his trading practices, Madoff showed them customer statements that seemed to disprove the front-running rumors. There is no indication that the young investigators ever tried to ascertain whether those trades were actually made.

The youthful SEC staffers were so dazzled to be at Madoff headquarters that they occasionally inquired about job openings at the company. "They would say, 'Can we get jobs—can we give you our résumés?' " Solomon said. "We'd say, 'Send them through to our offices.' A couple of them dropped résumés off." The secretary was appalled. "No wonder they never found anything," she said.

"I'd like to make a toast." Mitchell Horowitz tried to quiet down a ballroom filled with gray-haired couples chattering over their desserts.

It was a poignant occasion, the fiftieth anniversary party for his parents, Jerry and Doris Horowitz. At 76 years old, Jerry was in failing health, with several bouts of cancer behind him, and the couple wasn't celebrating at the fashionable Breakers hotel or Donald Trump's Mar-a-Lago estate tonight, but rather seventeen miles north at the Devonshire assisted-living complex in Palm Beach Gardens.

As retirement villages go, it didn't get much better. Located on the twenty-six-acre home of the Professional Golfers' Association, it provided resort-like amenities to its affluent retirees, from concierge service to beauty treatments. On this winter evening in 2004, the mansion-style clubhouse was decked out with gorgeous flowers and candles that gave the place a soft glow.

Jerry had the look of an aging linebacker, with a neck as wide as a tree trunk and an almost perfectly round, balding head. He looked dapper in his tuxedo, and Doris, wearing a long gown, was beaming. Their friends and family members took turns honoring them with words that brought some guests to tears. It was a magical evening.

Speaking into the microphone, Mitchell announced to the crowd that he wanted to take a moment to pay special homage to someone in the audience. The room fell quiet. "Tonight we drink to Bernie Madoff," he said. "We drink to Bernie because my parents are living in this beautiful place, and are able to enjoy this beautiful party because of him. We owe a special thanks to him."

He didn't have to explain what Madoff had done to allow his parents to live so well. Like Jerry and Doris, many in the room were living comfortable lives thanks to their profits from Madoff's investment fund. The audience stood and turned to Bernie, sitting at his table looking a little embarrassed. In unison, they raised their glasses and toasted him for bringing them so much joy. He acknowledged them with a modest smile and a small wave.

So many owed him their thanks, but no one in this room had as much to give thanks for as Jerry Horowitz.

It had been over forty years since Horowitz and Madoff had met at Saul Alpern's small accounting firm in Manhattan. Bernie was in his twenties and borrowing office space for his fledgling broker-dealer business. Horowitz was a part-time accountant, working with Frank Avellino and Michael Bienes.

Jerry had grown up in the shadow of Yankee Stadium in the Bronx and had a plainspoken, old-school quality to him. "He was a Mason and a simple guy who loved his wife and his kids," Bienes said. But he was also enormously bright, with a photographic memory and the ability to read an entire book in one night. His bookshelves contained thousands of titles he'd read over the years.

Several years older than Madoff, Horowitz struck up a friendship with him and started doing the taxes for his company. Alpern watched the relationship bloom with a wary eye, as Horowitz already had one foot out the door and was looking to start his own practice. Worried that he would lose Madoff's business, Alpern demanded his account back. "He went down to see Bernie and came back with his head in his hands," Bienes recalled. "Bernie said, 'No, Saul, I'm not going back with you. Jerry is my accountant, and Jerry stays my accountant.' "

Horowitz remained Madoff's one and only accountant for the next thirty years, doing the books for Bernie's mushrooming brokerage from the kitchen table of his home in Bellmore, Long Island. Accountants for brokerage houses have the responsibility to conduct impartial audits of their clients on behalf of investors, but theirs was far from an arm's-length relationship. Madoff was the client of a lifetime: he invested money for Horowitz and made him a millionaire.

Neil and Constance Friedman were feeling emotional this sparkling evening. They had been Jerry and Doris's best friends since the early 1960s. They were young couples with children back then, living on the second floor of a modest apartment building in Far Rockaway. It was thanks to Jerry that Neil and Connie met Bernie and were allowed into his investment fund in the early 1980s. Neil put $500 into his Madoff account, a move that changed their lives. The returns were so high that the Friedmans poured more money in over the years, until they'd invested every dime of their savings and retirement nest egg with Madoff—over $4 million in all.

Connie resisted the idea at first. Her family lived above a deli during the Second World War and was so poor that her mother could barely feed the family. "You didn't put money in stocks," she said. "You had cash—so if you needed it, it was there. You could buy food. You could pay your rent." All the talk about Madoff scared her, until Jerry assuaged her fears. Jerry was one of the smartest people she knew. He understood money better than almost anyone.

The profits the Friedmans reaped from their Madoff account were glorious; sometimes the returns approached 30 percent. The money was so bountiful that they retired in 1999 and moved to a large and airy two-story house in a gated community in Palm City, Florida.

Neil believed Madoff's claim that all of the fund's investments were cashed out every three months "to accommodate people who drew money out on a calendar quarter for a living." Neil thus made a practice of writing a letter to Madoff Investment Securities in New York every three months, requesting cash for Connie and him to live on. The redemptions were usually for around $40,000, with some extra cash withdrawn from time to time for dream vacations in places like Antarctica and Tahiti. The checks always came on time.

It was a good life. Connie's resistance to Madoff yielded to gratitude, and the Friedmans and the Horowitzes developed a tradition: each time they went out to dinner, they'd buy a bottle of wine and toast Bernie Madoff for the wonderful retirements he'd given them. Just as this crowd of a hundred was doing tonight.

By the night of the anniversary party, Jerry was in his twilight years. After his retirement, the Madoff account had stayed in the family, moving down a generation into the hands of Jerry's son-in-law, David Friehling. The Friedmans had attended his wedding ceremony at a lodge in the Catskills in the early 1980s. The dark and athletic-looking groom was a CPA, like his father-in-law. "Jerry was more personable than David," Neil said. But "David was young and also a very smart young man."

Horowitz brought him into the business in 1988, forming the accounting firm of Friehling and Horowitz. When Horowitz retired to Florida a few years later, Friehling had the company, and the Madoff account, all to himself.

Madoff Investment Securities was a multibillion-dollar business by then, handling 10 percent of the trades on the New York Stock Exchange and taking investment money from French banks and Arab sheiks. But Friehling, like his father-in-law, handled the entire account by himself, working out of his house in New York without so much as a secretary to help him. He finally splurged on a 13-by-18-foot storefront office on Main Street in New City, an upstate hamlet thirty miles north of Manhattan.

Each year, Friehling submitted annual certified audited

financial statements for Madoff's company to the SEC. These claimed that Friehling followed the Generally Accepted Accounting Principles in scrutinizing Madoff's finances. He certified that he had verified Madoff's assets and examined his bank accounts and that Madoff had actually purchased the securities his statements represented.

None of it was true. If Friehling had followed the required steps, he would have uncovered an unmistakable trail of clues to Madoff's scam.

It was easy to hide his actions. The American Institute of Certified Public Accountants regularly scrutinizes accounting firms to ensure they follow ethical business practices. But a firm can avoid the AICPA by declaring that it no longer performs audits. According to charges later filed against him, Friehling did just that, writing to the organization every year for fifteen years to claim that he didn't audit anyone. In reality, Friehling continued to submit annual audits for Madoff.

Like Horowitz, Friehling had his family fortune—$14 million—invested in Madoff's fund, and it was spitting out profits that were making his life increasingly comfortable. It wasn't in his interest to end the party.

If there was a single person at the Devonshire clubhouse this night who recognized the absurdity of one old man or his son-in-law doing all the auditing for a multibillion-dollar market-making and investment company, he or she never spoke up. If there were guests at this beautiful event who worried that Madoff was making millionaires out of the men who were supposed to be conducting independent audits of his firm, they kept their fears to themselves. No one wanted to cause a problem when it came to Bernie Madoff's miraculous investment machine. And so the men and women at the Devonshire clubhouse rose to their feet on this beautiful night and toasted Madoff and his longtime accountant. Years later, they would wonder how they could have been so blind.

The guests at the Horowitz event were people with varying financial sophistication. But the professionals running Madoff's billion-dollar feeder funds were accepting this surreal arrangement as well. The executives who ran Walter Noel's Fairfield Greenwich Group routinely assured investors of their superior due diligence of Madoff's operations. "Friehling and Horowitz, the independent auditors of BLM, conduct an annual report of the internal controls at BLM and have always provided a clean opinion," Fairfield's literature boasted.

On September 5, 2005, a client sent an email to Fairfield Greenwich inquiring whether it had thoroughly vetted the Friehling and Horowitz accounting firm. Fairfield's chief financial officer Dan Lipton wrote a response for Jeffrey Tucker to send back. Friehling and Horowitz, it stated, "have hundreds of clients and are well-respected in the local community."

But Noel's outfit had no contact with Friehling and Horowitz for sixteen years, according to the Massachusetts Secretary of the Commonwealth. In subsequent testimony before the Massachusetts Commonwealth Securities Division, Lipton stated that the firm was well-respected because a partner of Friehling and Horowitz—he couldn't remember his name—had told him so. That was also how he got his information that Friehling and Horowitz had hundreds of customers. Another Fairfield executive concluded that Friehling and Horowitz were independent auditors because their documents used the term *independent auditor* on the front cover of their reports.

In retrospect, Fairfield Greenwich's due diligence was surprisingly inept. But its officials seemed not to really want to know the truth about Friehling and Horowitz. On September 14, 2005, an official in the company's Bermuda office named Gordon McKenzie emailed a disturbing finding to the company's top brass. "It appears Friehling is the only employee" of Friehling and Horowitz, he stated. Jeffrey Tucker replied to his note with two words: "Thank you." And that was that.*

The fact that Madoff was employing just one accountant to handle the books for his multibillion-dollar operation was a pretty strong clue that something was a little off about his company. But government regulators missed it as well.

In October 2005, Harry Markopolos was optimistic that Bernie Madoff's moment of reckoning was approaching. Frank Casey had discovered that Madoff was attempting

---

*Fairfield Greenwich claims that its outside auditor, PriceWaterhouseCoopers, conducted "clean" audits of Fairfield Sentry and scrutinized Madoff. The Massachusetts Secretary of the Commonwealth has charged that in many instances its role consisted of merely interviewing Madoff about his practices.

to borrow money from European banks, leading him to speculate that the Ponzi scheme was running out of cash. On October 18, Markopolos emailed Casey and some others who were following his investigation into Madoff. "There's plenty of dry tinder laying next to the powder keg," he wrote. "Now all we're waiting for is the spark to set off the explosion."

He painted the doomsday scenario that a Madoff implosion would trigger, in which investors would flee the hedge fund market en masse, triggering a Wall Street meltdown. "Could the selling cascade into something much bigger, a la the '87 crash?" he asked. "It just might." The hedge fund industry would be changed forever, he predicted: "The fact that the world's largest hedge fund was operating unbeknownst to regulators ... will be a huge blow and lead to stricter regulation of hedge fund returns. Then the fact that FOF's [funds of funds, or groups of hedge funds] purposely ignored numerous red flags will call the entire FOF industry into question."

His prediction of a stock market crash was overblown, but the thrust of his email was otherwise prescient. Markopolos had once again seen what others had not.

Yet his point wasn't to warn his colleagues of impending doom; it was to advise them about how to profit from it. The subject heading of the email was "Possible Madoff Plays for Your Personal Accounts."

"Here's [how] to best play this in your PA," Markopolos wrote. He outlined a list of investing instructions geared to help his colleagues make money from the chaos that Madoff's collapse would trigger. He told them to buy options designed to generate a profit when the market fell. He also advised his friends to place bets that Oppenheimer, the parent company of Tremont Capital Management, a major Madoff feeder fund, would go bust: "Puts on Oppenheimer if they have put options listed on it would be a great trade as well. Unlike the indexes which may drop 25% max to perhaps 5% min, the owner of Tremont Capital will see its stock go to 0."

When a colleague emailed back to point out that he had his facts wrong and that Tremont's parent company could probably absorb whatever loss it took on Madoff, Markopolos was disappointed. "Darn It!" he wrote. "I was afraid of that." A dejected Markopolos vowed to find other ways to capitalize on the looming failure of Madoff and his feeder funds: "There's not a good individual equity put play with Madoff unless we can think of someone else with large exposure," he concluded ruefully.*

Markopolos may have been waging his crusade against Madoff for the American flag, as he later claimed. But he was also clearly out to make a buck on it.

It had been five years since Markopolos's first presentation to Grant Ward at the SEC's Boston office had fallen flat. But there were still staffers at that office who believed in his cause. A Boston staffer named Ed Manion had been his first

---

* An "equity put play" is a way Markopolos could have made money with options if a company that invested with Madoff failed.

ally, and he reached out to Markopolos and urged him to come back to speak with Mike Garrity, branch chief of the SEC's regional office.

In preparation for yet another meeting with the SEC, Markopolos decided to lay out his argument in a comprehensive memo. The October 25, 2005, document was a dense, rambling, twenty-one-page thicket of arguments, mathematical formulas, and Wall Street jargon, but Markopolos gave it a title anyone could understand: "The World's Largest Hedge Fund Is a Fraud."

"The fewer people who know who wrote this report the better," he stated in the introduction. "I am worried about the personal safety of myself and my family."

He argued that Madoff was either front-running or running the world's largest Ponzi scheme, which he deemed more likely. And he listed no fewer than twenty-nine red flags pointing to an illegal enterprise.

"Name one other prominent multi-billion dollar hedge fund that doesn't have outside, non-family professionals involved in the investment process," he wrote. "You can't because there aren't any."

Naturally, he brought up the federal reward money. "NOTE: I am pretty confident that Bernie Madoff is a Ponzi Scheme, but in the off chance he is front-running customer orders and his returns are real, then this case qualifies as insider-trading under the SEC's bounty program."

But the document was mostly dedicated to proving that the Wall Street legend was a con man:

Several equity derivatives professionals will all tell you that the split-strike conversion strategy that BM runs is an outright fraud and cannot possibly achieve 12% average annual returns with only 7 down months during a 14½ year time period. . . . It is mathematically impossible for a strategy using stock, individual stock call options and index put options to have such a low correlation to the market where its returns are supposedly being generated from. This makes no sense!

He took aim at Madoff's strict insistence on secrecy: "The investors that pony up the money don't know that Bernie Madoff is managing their money," he stated. He listed Walter Noel's Fairfield Sentry fund and Thierry de la Villehuchet's Access International Advisors as two feeder funds that were hiding Madoff's role.

The memo was wrong at times. "One London based hedge fund, fund of funds, representing Arab money, asked to send in a team of Big 4 accountants to conduct a performance audit during their planned due diligence," he reported. "They were told 'No, only Madoff's brother-in-law who owns his own accounting firm is allowed to audit performance for reasons of secrecy in order to keep Madoff's proprietary trading strategy secret so that nobody can copy it." Neither Jerry Horowitz nor David Friehling was related to Madoff. Yet even this erroneous account got the big picture right: Madoff kept outside auditors away from his records.

The memo was a devastating deconstruction of a scam. Its author had figured out the scheme by relying on publicly available information and easily obtainable client statements. Anyone who knew something about the hedge fund industry might have done the same thing.

"Bernie Madoff's returns aren't real," Markopolos concluded.

Harry met with Mike Garrity on October 25, 2005, and the SEC official was impressed. Markopolos walked away believing that someone in power had finally listened to him. "Unlike my disastrous May 2000 meeting with that office's Director of Enforcement, Attorney Grant Ward, I found Mr. Garrity to be interested and fully engaged in my telling of the scheme," he wrote his associates.

> Some of the derivatives math was difficult for him to understand, so I went to the white board and diagrammed out Madoff's purported strategy and its obvious failings until he understood it.
>
> Perhaps the most impressive thing about Mr. Garrity was his willingness to think outside of the box. He was able to imagine the impossibility of Madoff's returns and understand that Bernie Madoff's returns were too good to be true and this obviously concerned him. He told me that if Bernie Madoff were located within the New England region, he would have had an inspection team inside Bernie Madoff's operation the very next day.

But Madoff's company was not located in New England. Garrity was forced to pass the case off to his colleagues in Manhattan. And they weren't as enchanted with Harry Markopolos.

Meaghan Cheung served as branch chief of the SEC's New York enforcement division, which made her one of the government's top cops on Wall Street. Just four months earlier, her aggressive investigation into Adelphia Communications Corp. had resulted in the sentencing of its founder, John Rigas, to fifteen years in prison for looting the company of $100 million.

Excited that his case was headed for the big time, Markopolos emailed Cheung his opus on November 4 and followed up with phone calls in the following weeks. He later wrote that he'd learned some important facts from his calls.

"She's Korean," he wrote. And she was a lawyer, not a securities expert, which worried him. "The odds of an SEC branch chief and enforcement attorney understanding the trading strategy involved are less than zero," he stated. "What to me and the other derivatives professionals in the industry is obviously bogus, won't be intuitive to the SEC's staff." It was an enduring frustration for him; the regulators he pitched always seemed to be lawyers instead of MBAs or CPAs. He dismissed Adelphia as small potatoes compared to Madoff.

But overall, he came away feeling that the New York office was as excited by the case as Boston was. "Since I have to guess, here's what I think," he wrote to an associate. "New York read my report and said, 'holy shit, this is going to be big, we'd better assign a branch chief and enforcement lawyer to it ASAP.' "

A few days after their first conversation, he sent Cheung a note mentioning, among other things, the bounty. "Thanks," she responded.

Markopolos had envisioned Cheung's becoming a partner in his crusade, a member of his team. But it was wishful thinking. The branch chief saw Markopolos as a tipster to be debriefed, thanked, and shown the door. She refused to share with him any information about the SEC's internal operations, per agency policy. That meant he would be kept in the dark about the agency's response to his complaint. It wasn't the way he'd expected to be treated. And it wasn't long before he started to think of her as a useless bureaucrat. "She didn't give me anything concrete to assure me that my fears of the SEC ignoring my case for the 3rd time isn't a legitimate fear on my part," he wrote his colleagues. "I asked her in various ways to give me that assurance. She just said that they are looking into it."

Yet behind the scenes, the agency's investigators were intrigued.

The fall and winter of 2005 were unforgiving seasons at Madoff headquarters. First, Norman Levy died in September at the age of 93. Levy had been a father figure, business partner, and keeper of cash, and Bernie was moved to publish an uncharacteristically personal sentiment in a *New York Times* death notice. "My mentor and dear friend of 40 years. Your spirit and love of life have touched and changed all who knew you. You taught me so much. I'll cherish our relationship forever. Bernie Madoff."

Levy had lived a long life; Bernie worried that some members of his own family wouldn't be as lucky. Cancer had descended upon the Madoffs like a hailstorm. It began with Peter's bladder cancer in 2000, followed by a leukemia diagnosis for Bernie's 7-year-old grandniece, Ariel, in 2002. The following year, Andy's masseuse discovered a lump by his neck, which led to a diagnosis of mantle cell lymphoma, an extremely rare form of the disease. All three of them underwent treatment regimens that put their cases into remission.

But no one's ordeal monopolized the family's attention as much as Roger Madoff's. Peter's only son was a strikingly handsome and good-natured young man who had resisted the lure of the family business for a time, spending his twenties working as a reporter for *Bloomberg News* in Italy. One of his beats was covering automotive news, and he relished the days when he test-drove Ferraris through the Italian countryside. He finally joined his clan at the Lipstick Building in 2002. Soon afterward, he was diagnosed with leukemia.

Bernie and Peter summoned every weapon in their arsenal to help him. Bernie had company drivers chauffeur him to his medical appointments and provided him with a personal chef to prepare special meals when Roger became diabetic from his treatments. Peter spent countless days at his son's hospital bedside. His drivers would listen to him cry in the back seat of his car all the way back to the office.

In 2003, Shana, Roger's sister, made an extraordinary decision to save her brother's life. Doctors recommended that Roger undergo an allogeneic stem cell transplant to replace his white blood cells and allow his body to build an entirely new, cancer-free immune system. As his sister, Shana was a candidate donor. But she was pregnant at the time, and the transplant posed a grave danger to the fetus. Faced with the wrenching choice of having her baby or saving her brother's life, she ended the pregnancy and donated her stem cells to him.

The transplant didn't work as intended. Shana's white blood cells were rejected by Roger's body and he suffered through agonizing complications. By Christmas of 2005, the boyish young man had lost all his hair and was growing more emaciated by the day. A journalist at heart, he chronicled his experiences, hoping to publish his account of his battle against cancer in a book he titled *Leukemia for Chickens: One Wimp's Tale about Living through Cancer.*

With the office tense over Roger's deepening crisis, Bernie received news that dramatically compounded the pressure: the SEC was preparing to launch an enforcement case on his company in response to Harry Markopolos's complaint. It was a far more intensive investigation than the agency had ever conducted into Madoff's dealings. Investigators planned to scrutinize his operation as well as that of his largest feeder fund, Fairfield Sentry. The probe was clearly aimed at Madoff's investment business. It was a mortal threat to his criminal enterprise.

Madoff panicked. In December, he held a tense conference call with Amit Vijayvergiya, Fairfield's Bermuda-based chief risk officer, and Mark McKeefrey, its general counsel and chief operating officer. "Obviously, first of all, this conversation never took place, Mark, okay?" Madoff began.

"Yes, of course," Vijayvergiya said. Madoff was unaware, however, that the conversation was being taped, which Fairfield did on occasion.

It was well known that Fairfield Sentry was a pipeline to Madoff, but he seemed desperate to obscure his role. His investment advisory business wasn't registered with the SEC—it would have invited too much scrutiny—so he asked Fairfield to back up his story. The company line, he said, was that he was merely executing orders on their instructions: "In the past, if we've ever been asked about what our role is with any of these types of funds, it has always been that we are the executing broker for these transactions and that you use a proprietary trading model that we—that is ours that basically sets the—that, you know, has certain parameters built into it which have been approved by you and then that's part of the trading directive that you've seen."

"Right," said Vijayvergiya.

And so it went throughout the call, the big boss nervous and stammering as he urged them to minimize his relationship with them to the SEC. "I'm not telling you to conceal anything," he insisted. "I'm telling you, you know, that there are things that you don't—one of the problems we've had in the past is people go out and they—they, you know, even—I'm talking about like with you where guys write things in a document or say things which is not really—which is not a hundred percent the case."

Madoff coached McKeefrey and Vijayvergiya to say the words he uttered every time someone asked him to explain his mysterious split-strike strategy: "If they ask . . . do you know how Madoff decides when he's going to go in the market and out of the market, which is a question people always ask me, not only—and I'd say, you know, I'm not going to share that information with you. There's all sorts of—we have—you know, obviously a lot of black box technology, momentum models and all sorts of things that tell us when to get in and out of the market."

His anxiety was obvious, as when he excused himself to answer another call and then came back on the line. "I'm sorry," he told them. "If I get any more solicitations for charity I'm going to kill myself."

He concluded the chat with some final advice to the Fairfield executives. "You don't want them to think that you're concerned about anything," he said. "With them you should—you're best off if you just be, you know, casual."

"We're trying," McKeefrey said. "We're trying to be cool and to just cooperate and get it over with and get them out of here."*

It was a scene that would have given Harry Markopolos great pleasure: Bernie Madoff was squirming.

---

*Fairfield Greenwich claims it asked permission from the SEC to speak with Madoff before the conversation took place and subsequently informed the agency about it afterward.

The walls seemed to be closing in on him. The SEC's case opening report on January 24, 2006 listed its mission as determining whether Bernard L. Madoff was running a Ponzi scheme. The SEC staff stated that it had already concluded that Madoff had misled its investigators in the past about both his role in managing the Fairfield Sentry fund and the nature of the strategy he was using for it, his elaborate coaching efforts notwithstanding.

The SEC staff gathered information over the next few months as Madoff sweated it out. Then, on April 15, Roger Madoff died.

The family was devastated. Peter, badly shaken by his son's death, had never been very religious, but he retreated into the sanctum of religion afterward, attending synagogue each morning before work.

Bernie had been living a secret and deeply immoral life for several decades by then, and his nephew's death didn't spark a similar spiritual awakening. Indeed, just four weeks after Roger's death, he sinned again. On May 19, he testified under oath to SEC investigators and brazenly lied to them.

He appeared nervous to some people in the office that day, but when the SEC attorneys asked him about his investment practices, he delivered a seamless performance. He outlined the mechanics of stock and options trades he knew had never taken place, and he nonchalantly described the parties he traded with on a daily basis, people who in reality didn't exist.

"You mentioned that there was a group of dealers to whom you put out this indication of interest each time," his interrogator said. "Generally, with how many of them do you end up trading in each execution?"

"Within the basket, we're probably interacting with 40, close to 50," Madoff replied.

"That's for equities and options?"

"Equities. Options is a dozen."

"Who are the counterparties to the options contracts?"

"They're basically European banks."

"So how does the time frame for the options trading relate to the time frame for the equities trading?"

"First we're putting the equity basket on, and then we're putting the options on after the equity basket is complete, so the options are being done basically in the morning typically between 8:00 and 9:00 a.m."

He spoke with impressive ease and specificity, the same traits that made Michael Ocrant second-guess himself after he interviewed Madoff for *MARHedge*. Madoff also showed the investigators stock transaction records and investor records that seemed to jibe with the information they'd obtained from investors.

The SEC didn't release its findings for another year and a half. In that time, Harry Markopolos wrote off Meaghan Cheung and then brought his findings to the *Wall Street Journal*. Once again, his hopes rose after a conversation with John Wilke, a reporter. "They're going to want to dig deep, real deep and it looks like they're going to investigate Bernie Madoff's entire 40 year career looking for dirt," he wrote a colleague. With characteristic overkill, Markopolos bombarded the reporter with documents, contacts, questions for him to ask, and even ideas for follow-up stories.

But the *Journal* never did investigate Bernie Madoff. Markopolos believed it was the fault of "senior editors" at the *Journal* who "respected and feared" Madoff. Wilke never explained his reasons.

Markopolos was forever finding new people to help him expose Madoff, only to have his hopes dashed in the end. Either he was finding the wrong people or they weren't finding him very credible.

He finally accepted that he would never see a nickel of bounty money for his crusade. "The only thing I stand to gain (probably) is lots of free publicity and perhaps invitations to speak at hedge fund conferences," he wrote Frank Casey in May 2006. "I doubt there's any money in the Madoff case. If it is a Ponzi, the SEC will be more worried about giving the suckers pennies on the dollar back and will likely screw me out of the reward money."

He took a few more stabs at bringing new information to the SEC's attention, but grew increasingly dejected. "I'm not sure sending the SEC anything would help those morons solve the case," he wrote an associate in August 2007. "They're so lame, I'll bet they don't even catch colds in the winter." He reserved a special resentment for the New York SEC chief. "Every phone call to Meaghan Cheung made me feel diminished as a person," he wrote.

On November 21, 2007, the SEC Enforcement Division closed the book on its investigation into Bernie Madoff. "The staff found no evidence of fraud," the closing document stated. It concluded that Madoff had violated SEC requirements by failing to register as an investment advisor, and that Fairfield Greenwich Group hadn't adequately disclosed that Madoff was doing the investing for the fund. As a result, Madoff registered as an investment advisor and Fairfield Greenwich revised its disclosure statements. "Those violations were not so serious as to warrant an enforcement action," the SEC staff stated.

Two years later, a reporter for the *New York Post* would corner Meaghan Cheung outside her Manhattan apartment building and ask her about the case. She said she was shocked to learn that Madoff had indeed been running a Ponzi scheme all those years. Choking back tears, she insisted there wasn't much more she could have done to uncover it. "If someone provides you with the wrong set of books," she said, "I don't know how you find the real books."

Bernie Madoff was off the hook. Harry Markopolos tried one last time, in the spring of 2008, to interest an SEC official in Washington to look into the case. He mailed him all his findings but never heard back from him either. "I tried calling back a few times but never got through and gave up," he said. His fight was over.

10

The Fall of Bernie Madoff

On the morning of Friday, January 18, 2008, President George W. Bush stepped to a microphone inside the Roosevelt Room of the White House and offered his understated description of a deteriorating situation. "We're in the midst of a challenging period," he said.

~~Markets around~~ the globe were plunging as fears of a recession spread. The Dow had fallen 300 points the previous day after Merrill Lynch reported a $10 billion quarterly loss, a result of the exploding subprime mortgage crisis. The rapidly unfolding events prompted Bush to rush out an emergency $145 billion economic stimulus bill. "The fear is spreading," reported the *New York Times*.

At about the same time, Ruth Madoff was getting her hair done at the Frédéric Fekkai salon in Palm Beach. It was the start of a fairly typical weekend for her. Strolling under a hot sun and a perfect blue sky, she dropped in at stores along Worth Avenue and elsewhere, picking up $1,200 in clothing at the European boutique Diane Firsten on Saturday and plunking down $530 at Polo and $260 at Williams-Sonoma on Sunday.

She and Bernie were never in one place very long. They'd been in Manhattan that week and in Paris the weekend before, where she window-shopped and picked up $4,000 in clothing from Armani, Jil Sander, and Marni. It was all charged to her company AMEX card back at Bernard Madoff Investment Securities, which paid for it with stolen money.

As the economic free-fall continued and investment houses began to teeter, losses on Wall Street began to pile up and anxiety swept through middle-class America. But the souring economy in the first months of 2008 didn't seem to cramp Madoff's business or his lifestyle. His investors were delighted to see profits rolling in. He had lowered his rate of return drastically, to about 4.5 percent, but that was Nirvana compared to the wreckage taking place elsewhere.

While investment banks were laying off workers, Madoff's employees were enjoying job security. And his family members and closest aides were living a glorious life. On the same day the company paid for Ruth to have her hair done at Frédéric Fekkai, Andy spent $1,126 for dinner at Zagat's top-rated Manhattan restaurant, Per Se, in the Time Warner Center. That week, Shana charged the company for a flight to Cancun, Mexico, for her husband and her; Frank DiPascali bought a trip to the Bahamas; and Jodi Crupi, one of

DiPascali's top lieutenants on the seventeenth floor, booked a stay in Las Vegas. Just a week earlier, Crupi had dropped a thousand dollars on a dinner at Il Mulino, one of Manhattan's best Italian restaurants. Soon after, the company paid for Mark and Andy to relax at a luxury ski resort in Jackson Hole, Wyoming.

Ruth usually spent just one day a week in her eighteenth-floor office going through company bills and handling Bernie's personal expenses. Most of her time was spent planning her trips with Bernie, filling their social calendar, and writing checks to charities. She was forever tending to her appearance, exercising to keep her petite body trim, having her hair colored in one city and styled in another. She doted on the people who kept her looking chic and attractive; only Giselle at Pierre Michel on East Fifty-seventh Street was allowed to color her hair its Soft Baby Blonde color. Every six weeks, she got foil highlights at the large and opulent salon, where the smell of hairspray mixed with the chatter of her fellow Upper East Side ladies of leisure.

When Ruth and Bernie weren't flying private jets across the Atlantic or enjoying the smell of salt air in one of their yachts, they were home inside their 10,000-square-foot Lexington Avenue penthouse. Two writers Ruth invited up to the apartment to plan a scrapbook for Bernie's birthday in 2003 dubbed the furnishings "Queens High Baroque." "Gold sconces lined the stenciled wallpaper, a Napoleonic-style desk stood to the side, and the Greek and Egyptian statues vied with each other to set a mood of antique decorum," Michael

Skakun and Ken Libo wrote. "Arabesque-styled Central Asian rugs beguiled our vision with looping patterns and impressive symmetries, further softening our footfalls."

Bernie and Ruth were living out the materialistic fantasies of Laurelton's second-generation immigrant families. They seemed enormously proud of, and entitled to, their success. But they were living on other people's money. It was a sin for which Bernie showed no apparent remorse, and Ruth's friends saw no sign that she felt the slightest bit self-conscious or guilt-ridden about her privileged lifestyle.

Bernie had to be worrying about the economy: over the course of a March weekend, Wall Street giant Bear Stearns collapsed, sending the Treasury secretary and Federal Reserve officials into a frenzied round-the-clock effort to patch together a JPMorgan Chase takeover, greased with $30 billion in taxpayer backing. Madoff's investment fund was as vulnerable to panic selling as any other, but he didn't seem to smell the trouble that was coming. Some investors were actually flocking to him because his was seen as a safe and stable fund.

He was turning 70, and he marked the milestone with a joyous celebration. In May 2008, he and his clan traveled to Cabo San Lucas, a resort city in Mexico, for an industry golf tournament. About a dozen employees and family members held a birthday party for him on the beach, wearing sweatshirts with "BLM 1938," his initials and year of birth, emblazoned on them. He was so happy he sang along to "Sweet Caroline" by his favorite crooner, Neil Diamond. He was surrounded by people he loved, celebrating a career marked by phenomenal success. It may have been one of his last happy moments.

Madoff's investment machine was humming along and he was still turning down investors. A month after the party in Cabo San Lucas, he got a call from Stephen Richards. The souring economy hadn't prompted Bernie's old friend from Laurelton to get cold feet about his investment. On the contrary, Richards was calling to ask whether Madoff would accept another client. His brother-in-law had recently called him. "Do me a favor," he asked Stephen. "Call Bernie. Ask him if he'll take my 401(k)." He had a million and a half in it and had heard how Madoff made people like him fantastically wealthy.

Richards sounded sheepish on the phone as he asked Madoff to take his brother-in-law's money. Madoff was always nice to people from the old days, but he gently declined. The investment was too small for him, he said, and he had no place to put it. When his brother-in-law called Madoff himself to plead his case, he turned him down as well.

Yet if it was still business as usual at Madoff headquarters, anxiety was beginning to grip the offices of his biggest feeder fund. The meltdown on Wall Street had investors at Fairfield Greenwich nervous about money, and they were beginning to listen more closely to the rumors about Bernie Madoff.

In May, officials at a Swiss investment advisory firm named Unigestion started emailing questions to Amit Vijay-vergiya at Fairfield Greenwich's Bermuda office. Where was the proof that Bernie Madoff was actually making trades? Where were the assets he claimed to be holding? "We think that we have a counterparty risk as all the assets are held at Bernard Madoff Securities and it is very difficult to find infos on it, audit report, etc.," a Unigestion principal complained in a July 29 email.

Despite Vijayvergiya's assurances to the contrary, he and his colleagues seemed to have only a vague idea of how Madoff's investment operation worked. Walter Noel's lieutenants were caught between investors demanding answers and an investing wizard who wouldn't cough them up. "Unfortunately," Vijayvergiya wrote his colleagues in an internal email, "there are certain aspects of BLM's operations that remain unclear and although we are attempting to obtain responses from Bernie Madoff . . . this process could take some time."

But Unigestion wouldn't wait. It pulled $75 million in holdings the next month. That was a drop in the bucket compared to what came next.

Years earlier, JPMorgan Chase in London had invested $250 million of its own money in Fairfield Sentry and its companion fund, Fairfield Sigma. For years, the bank collected handsome profits from Bernie Madoff. But with market conditions eroding, the company conducted a review of all its hedge fund holdings, and in August decided that Madoff represented a risk. The due diligence questions started flying, and officials at Fairfield Greenwich were once again stuck for answers. Chase yanked its quarter-billion-dollar investment in August 2008. Bank officials said they became concerned about the "lack of transparency to some questions we posed as part of our review," a spokesman explained.

Noel's aides grasped that events were threatening to capsize their multibillion-dollar money machine. Yet as the world was changing and parts of it were collapsing, Bernie Madoff was acting like a man without a care in the world.

The economic disaster unfolding in 2008 reached a climax the week of September 15. On that Monday morning, Americans awoke to discover that the country's financial system was breaking down.

Federal officials had spent yet another grueling weekend trying to forestall the implosion of two great financial houses. Like Bear Stearns before it, Lehman Brothers was struggling under the weight of billions of dollars in worthless derivatives tied to subprime mortgages. But this time, Treasury Secretary Hank Paulson and Federal Reserve Chairman Ben Bernanke balked at a federal bailout and decided to let the market sort itself out. It was a catastrophic decision. Just after midnight, Lehman announced its decision to file for bankruptcy—the largest filing in history up to then. Meanwhile, Merrill Lynch, the legendary Wall Street powerhouse, was also teetering at the edge of bankruptcy. To save it, Bernanke and Paulson brokered a shotgun marriage with Bank of America.

The Dow Jones plunged 504 points, the largest loss since the September 11 attacks. NASDAQ stocks plunged by almost 4 percent.

The next day, the Bush administration announced an unprecedented $85 billion bailout of AIG, the insurance giant that served as a crucial artery in the country's financial bloodstream. Financial institutions whose securities it had insured were toppling, and AIG's own investment arm was imploding from the derivatives mess that had broken the back of Lehman, Bear Stearns, and Merrill.

Pillars of the nation's economy were disintegrating. Millions of Americans watched the value of their investments nosedive. The shutdown of the credit markets halted the flow of money coursing through Wall Street and brought activity in many companies to an eerie dead halt, drying up bank lending. It seemed as if America was going broke in the course of a few days.

At the end of the week President Bush stepped in and tried to put out the fire. He announced a historic $700 billion package to buy troubled mortgages and free up credit. It was a remarkable concession from a free-market administration. "It's a big package because it's a big problem," the president said.

One of the only exceptions to the halt in market activity could be found at Bernie Madoff's investment operation inside the Lipstick Building.

Investors spooked by the chaos in the markets continued to migrate to him as the only port left standing in the storm. Madoff's fund was known for its stability. His solid 4.5 percent return was looking terrific compared to the carnage everywhere else.

René-Thierry Magon de la Villehuchet was one such investor. Battered by the decline of his other holdings, he decided to plow more of his money into Madoff, bringing his total stake to roughly $1.4 billion. The number included almost every dime of his $50 million fortune.

But Madoff's success was starting to backfire on him. Two days after the president's announcement, Richard Landsberger, a member of Fairfield Greenwich's executive committee, sent a message to Jeffrey Tucker and fellow committee members. He simply couldn't figure out how, in the midst of a credit crisis, Madoff was able to find trading partners willing to cover all the options he was supposedly using to execute his split-strike strategy. "How do the counterparties at the banks provide enough liquidity to allow the options to be put on?" Landsberger asked. "My understanding is dealing desks are currently doing almost everything on an order or best efforts basis—and not providing risk capital to anyone. Can we get some clarity from BLM on how he sees the markets and liquidity from his counterparties on the options?"

Two days later, his request took on more urgency. "I don't want to beat a dead horse, but this illiquidity exists in the bund, jbg and usd bond markets as well," he wrote, referring to the German, Japanese, and U.S. markets. "Does anyone in this email not think we should speak to BLM asap and get some color from him on how we are getting option liquidity . . . ?" The implication was that Madoff was up to something strange.

What followed was a chaotic effort by Fairfield executives

to figure out who Madoff's trading partners were, while at the same time trying to make their clients believe they already knew.

Frank DiPascali assured Vijayvergiya that Madoff was dealing with twenty brokers and international banks, "primarily Europeans for options," he added. But that's all he'd say; he offered no names, citing competitive reasons. Vijayvergiya's bosses and partners grew increasingly nervous. Andrew Smith, a Fairfield partner, wrote, "In times like these [it] would be great if we could get some clarity on who the 20 non-US options counterparties are an[d] the liquidity in the mktplace just for our comfort, not to tell the clients, so we can tell them we know/monitor."

On October 2, Noel, Tucker, and McKeefrey trekked to the Lipstick Building to confront the Wizard of Oz. Madoff was polite and generous with his time as he blew off their concerns. Fairfield had prepared a list of questions for him, and the answers they recorded reflected his dismissive attitude. When the Fairfield executives pressed him to reveal who was buying his options, the answer one of the participants penned was "BLM will not disclose the names of the c/p's [counterparties] for 'obvious reasons' (i.e. confidentiality)." Asked for the names of his employees who were executing the split-strike formula, he wrote, "No names given." Asked who placed the trade orders for the strategy, Madoff replied, "Traders, under the direction of supervisors."

His visitors were either too polite or too intimidated to press him further. After twenty years, Walter Noel was kept in the dark about the investment operation handling billions of dollars of his clients' money. And though his clients were now hounding Noel for answers, Madoff refused to help him.

Madoff wasn't simply being arrogant. He couldn't give Noel names because there were none to give; there was no one executing a split-strike strategy and no counterparties buying options for it. The twenty options dealers in Europe that Frank DiPascali spoke of didn't exist.

Noel and his aides may have been too cowed to force the issue, but it was Madoff who would ultimately suffer. While de la Villehuchet and other clients were moving their money into his fund, many more of them, suffering from the market meltdown, were lining up to take their money out.

In November, the employees lingering around their desks on the nineteenth floor started to gossip about Bernie's behavior. He seemed subdued, as if he'd lost his spirit. His obnoxious sarcasm was missing. He was locking himself in his office all day, and people could see through the glass walls that he was spending all his time hunched over his computer, poring through numbers. When Peter's secretary, Elaine Solomon, or Bernie's secretary, Eleanor Squillari, knocked on his door to tell him about an important phone call, he waved them away, shouting, "Tell him I'll call him back."

"What's with Bernie?" a few people asked Eleanor. "I don't know," she told them. "He's like in a coma."

It was more like Madoff was starting to sweat. Every

Ponzi schemer fears the moment when he no longer has enough money to fund people's withdrawal orders, and Madoff could see that moment in the distance. He'd reported to the SEC that his investment advisory operation had $17.1 billion in assets, but it was a lie. Madoff likely had less than a billion dollars left in his bank account by then, and customers hit hard by the financial crisis were lining up to redeem their investments.

Desperate for cash, he decided to announce the formation of a new billion-dollar investment fund. He reached back into his playbook and pulled out his tried-and-true strategy of playing up its exclusivity, decreeing that he would open its velvet ropes only to an elite handful of high-worth investors. Then, for the first time in forty years, the mysterious investment king picked up the phone and started dialing for dollars.

On November 24, he and the family were preparing to head to Palm Beach for the Thanksgiving holiday. But first he needed to attend to some important business. A potential client was coming into the office to chat about his new investment fund. Ken Langone, the crusty, hawk-faced cofounder of Home Depot, should have hit it off right away with Madoff. They were both Wall Street elder statesmen, wealthy beyond imagination, practically the same age, and proud of rising from modest beginnings. Langone had heard about Madoff's legendary investment prowess for years.

The two walked together through the corridors as Madoff showed off the artwork on his walls. But when Langone, his business partner, Steve Holzman, and Langone's son Bruce sat down at a conference table, the pitch went almost immediately awry.

Pushing the ultra-exclusive nature of his new fund, Madoff told Langone that he wanted just five investors in it. He explained that it would employ the split-strike conversion formula that had performed so well for so many in the past. Then he added a sweetener. "This fund is going to do better than my existing investors [do]," he said. "This is an unusual opportunity." The only condition was that he'd need Langone's money soon.

Langone was appalled. Madoff was betraying his longtime investors by offering higher returns to his newer ones. "I thought how I would feel if I was already in and he was doing this with me," Langone recalled. Madoff had overplayed his hand.

Things didn't improve when Holzman started peppering Madoff with questions about the mechanics of the split-strike strategy. "I can't talk about this," he replied, his stock response. It was proprietary information.

When the meeting ended, Langone rode down to the lobby with his son and partner. "That guy does nothing for me," he told them.

Madoff needed to find an investor willing to part with his money in the middle of the worst financial crisis since the Depression. Scouring his Rolodex, he hit on the one person he'd always been able to count on.

Nine months after they had sat together at his ninety-fifth

birthday party at Club Colette, Bernie Madoff called Carl Shapiro. He was an anxious son in a jam turning to his father. But instead of asking for a few bucks to fix a car he'd dented, Madoff hit up Shapiro for a quarter of a billion dollars. It would have been a laughable request coming from someone else, but Bernie and Carl were family. The old man trusted him like the son he never had. Like a son with a dented car, Madoff told Shapiro he'd pay him back soon. Shapiro agreed to send him the money.

It was an unconscionable act. With his Ponzi scheme on the verge of collapse, Madoff knew it was likely Shapiro would never see his money again. Carl would have done anything for him, and Bernie preyed on that loyalty to help salvage his criminal enterprise. It was a tragic betrayal.

In the days that followed, Madoff called the Shapiro residence incessantly to check on the money. "I didn't get it. It hasn't come yet," he'd tell Shapiro, his anxiety rising with each call. "Are you sure you sent it?" The cash eventually arrived, but things were unraveling so quickly that it bought him a few days at best. The pressure on Madoff was mounting.

His assistants knew something was wrong but couldn't figure out what it was. Bernie wasn't someone who shared his problems with people. Instead, he simply withdrew. "He was very quiet," said Elaine Solomon. "He was preoccupied all the time." It was hard to read his face; his expression was dour on a good day, smiling didn't come naturally, and friends had long grown used to the look of worry he wore.

But the panic he was trying to keep bottled up occasionally seeped out.

Two days before Thanksgiving, Madoff made an unusual request of his wife. Eyeing disaster, he asked Ruth to transfer $5.5 million from her account at Cohmad Securities, a brokerage firm partially owned by Madoff, into her personal bank account. She followed his instructions, extracting the stolen money for safekeeping.

In the days that followed, he grew obsessed with his health. "I'm going to the doctor to get my blood pressure checked," he told his aides more than once. He eventually came to work with his own blood pressure monitor.

Then his back started to hurt. It wasn't unusual for Bernie to take naps on his office couch, but now he was lying down on the floor of his glass-enclosed office to give his aching muscles a hard, flat surface. Quizzical employees walked by his office and saw their boss splayed out like a corpse. "Is Bernie all right?" one staffer asked Eleanor Squillari. "Well, he's not dead," she replied. "But he's not all right."

Madoff was planning for the worst. On December 1, he placed a call to his longtime attorney, Ira Sorkin, a specialist in white-collar criminal defense. He told Sorkin he needed to see him but couldn't do so for another two weeks. They made an appointment for December 12.

Madoff's desperate search for cash led him to the president of a Bronx fuel oil company. Martin Rosenman's grandfather had founded the firm in 1934 as an ice delivery firm. Rosenman had heard great things about Madoff, and on December

3, he agreed to invest $10 million of his family's money with him. Even in his desperate state, Madoff played hard to get. His fund was closed for another month, he told Rosenman, but he could wire the money immediately and Madoff would hold it until the first of the year. Four days later, according to Rosenman, Jodi Crupi on the seventeenth floor sent him a fax instructing him to wire the money to Madoff's account at JPMorgan Chase. Four days after that, he received a confirmation from Madoff Securities stating that his money had been invested in Treasury bills. Rosenman tried repeatedly to find the transaction online and couldn't locate it. He called Crupi, but she never returned his call, he said. He'd been conned.

He was just the latest of many. The November statements being churned out on the seventeenth floor reported that Madoff's clients had assets worth $65 billion. The true number was closer to zero.

Despite Madoff's frantic fundraising, there was no way to compensate for the hemorrhage. Clients had pulled $7 billion from their Madoff accounts. His worst nightmare was coming true.

On December 8, Madoff exploded in anger.

He got Jeffrey Tucker on the phone and started railing about the flood of redemption orders coming from Fairfield Greenwich. If Fairfield couldn't replace the money its clients were pulling out, he'd close its account altogether, he warned. His traders were already "tired of dealing with these hedge funds," he said. "There are plenty of institutions who can replace the money." He'd received offers already but had stuck by Fairfield out of loyalty, he said.

Tucker was shaken. He sent a code-red email to Walter Noel and the company's executive committee. "Just got off the phone with a very angry Bernie," Tucker wrote, "who said if we can't replace the redemptions for 12/31 he is going to close the account. . . . Not sure of our next step but we best talk. I think he is sincere."

Fairfield Greenwich was Madoff's best customer, with almost $8 billion in his fund (or so its owners thought), and yet he'd berated Tucker as if he were an incompetent employee. The fact that Madoff could make Tucker tremble and not the other way around was a remarkable testament to the bizarre balance of power.

Noel and Tucker were in a bind. They had already been working feverishly to placate Madoff, forming two new feeder funds, BBH Emerald and Greenwich Emerald Funds, and raising millions for them. Whether they'd told their investors that their cash was going to prop up Bernie Madoff's sinking fund was another matter.

Noel and Tucker even dug into their own pockets. "We tried to help stem things," Noel said later. "We thought, well, we can help him a bit if we give him some more money." But it was a drop in the bucket compared to Madoff's needs. Nervous about being cut off, Tucker penned a long, groveling letter to Madoff, as if Fairfield had done something wrong. "We apologize for failing to keep you informed of pending redemptions in a timely manner," Tucker wrote. "Our firm is very dependent on its relationship with you. You are our most important business partner and an immensely respected friend. As a firm, we are prepared to commit to dedicating ourselves exclusively to Sentry and Emerald. Throughout 2009, we will engage in no other fund-raising initiatives. Our mission is to remain in business with you and to keep your trust." He ticked off a number of steps intended to get back into Madoff's good graces.

It was a final bow of fealty to the man Bernie Madoff had once been. But by the time the letter arrived, hand-delivered to the Lipstick Building on the morning of December 10, 2008, Madoff and his business were going up in flames.

Bernie loved his family. His face lit up when Ruth walked into a room. He spent his days and many of his evenings with his sons, lavished seven-figure salaries on them, and begged out of dinners and canceled plans to spend time with their families. He and his brother, Peter, suffered grievously when their sons developed cancer.

He may well have kept Peter, Mark, and Andy away from the seventeenth floor to shield them from his criminal enterprise. But in his irascible moments, Bernie often described his sons as spoiled and undeserving of the positions he'd given them, and he seemed to view his younger brother as something of an egghead. Madoff may have frozen them out of the investment advisory operation because he felt they weren't competent or tough enough to help run it. He had chosen Frank DiPascali, who reminded people of a loan shark, for a reason.

But if his family members were pushed away from the investment business, that didn't mean they were cut off from it. The clues to Bernie's crime were all around them. Peter was the chief compliance officer; it was his job to deal with the SEC, which investigated Madoff at least five times over twenty years. Andy and Mark ran the trading operation, which was subsidized by Madoff's investment advisory business. They were also directors of the London office, the hub of his money-laundering operation. The family members had also read the *MARHedge* and *Barron's* stories, with their implications that crimes were being committed downstairs.

The boys and their uncle could probably have uncovered Bernie's crimes if they'd followed any of the dozens of obvious clues. But that would have taken a desire to learn the truth.

They would soon know it all. By the second week of December, the house of cards had collapsed. Madoff was almost out of money. He knew that once he was unable to honor redemption orders, angry clients would call the SEC and that would be the end of things.

A calm washed over him. It was time to reveal his secret.

Sitting alone in his glass sanctum, he plotted the rollout of his surrender. He would disburse the last of the investment fund's cash, somewhere between $200 and $300 million, to a select group of family members, employees, and friends and turn himself in the following week.

On Tuesday, December 9, the rollout began. He asked

Mark for a list of annual bonuses to pay out even though it was two months early for them. The move made little sense to his son and left him unsettled. He'd been worried about his father—he'd seemed depressed in the previous few weeks—and the news left him wondering if the company was in trouble. He decided to speak to Andy about it.

Bernie then took a deep breath and asked Peter if he had a moment to talk. Then he closed the door behind him and confessed his crime.

Peter had weathered the loss of a son and battled cancer himself, trials that had driven him into the arms of God. Now another unspeakable tragedy had come through his door. His older brother stood before him and announced that he was a crook and had spent his entire lifetime lying to him. Peter's career was over. Much of his wealth would vanish. He could end up in jail himself.

The implications would take time to absorb. Peter was a lawyer and knew he needed to turn his brother in. But Bernie had a plan he wanted to follow and it didn't involve going to the authorities yet. Peter deferred to his brother, as usual.

It was raining on the morning of December 10 as the employees of Bernard Madoff Securities LLC started to straggle in to work. It was the day of the company Christmas party, a reliably grand Madoff production. While other companies were closing or laying off workers, Madoff shelled out $30,000 for the event. It was further proof to his employees that they were leading a blessed existence, that their jobs were safe. They had no way of knowing that all but a few of them would be out on the street by the end of the week.

Eleanor Squillari thought it was strange that Bernie hadn't scheduled any meetings or phone calls that day. Her curiosity was further piqued when Ruth walked in and rushed by her without saying hello. It was only later that she learned that Ruth had come to withdraw another $10 million from her account at Cohmad.

As Squillari and Elaine Solomon made last-minute arrangements for the evening's party, Andy and Mark walked over from the trading floor, looking anxious. They entered Bernie's office; Peter joined them. They were awfully grim, Solomon noticed. But the Madoff men were always moody.

The sons confronted their father about the early bonuses. They demanded that he level with them about what was going on.

Bernie never had a problem lying when confronted. The company was having a good season, he said. It was a good time to share the profits.

But they didn't buy it. He'd been acting strangely over the past few weeks, and now he had ordered up a bizarre payment schedule out of the blue. Was the company in trouble?

They were forcing a confession before he'd had a chance to plan one. When cornered, the old Bernie Madoff would have instinctively turned on the charm and put on a convincing show. But this time he couldn't hold up the veneer. He looked ashen. "I'm not sure I'm going to be able to hold it together," he said. He wanted to reconvene at his apartment.

They could all be seen through the glass walls of Bernie's office. Eleanor later wrote in *Vanity Fair* that Peter "looked as if the air had been sucked out of him." Without thinking twice, she ambled into the office to drop off the mail. "Bernie and his sons stood up, startled, and stared at me," she recalled.

When the meeting ended, Andy and Mark helped Bernie put on his coat and they all walked out of his office together.

"And where are you going?" Squillari asked Madoff.

"I'm going out," he said.

Mark turned to the secretary and whispered, "We're going Christmas shopping."

Clive Brown and the other company drivers were on the seventeenth floor idling in "the Cage," an ancient Wall Street term for the area where paper trading tickets came in for processing. In modern times it was the room that handled wire transfers.

Eleanor Squillari called down and asked Brown to take Bernie to his apartment on Sixty-fourth Street and Lexington Avenue. It was about 9:30 a.m. He'd be ready to leave in ninety minutes.

Brown left the Lipstick Building and picked up the Mercedes at the garage. On the way back, he got caught in traffic. His cell phone started ringing. It was Andy, wondering where he was.

When Brown pulled up, Bernie, Andy, and Mark were already waiting for him. They squeezed into the car together.

Brown proceeded west along Fifty-third Street to Park Avenue, then turned north to Sixty-fourth Street.

Bernie broke the silence. "How are the kids?" he asked them.

The sons and their father kept the conversation light. Andy said that his daughter, a private school student, had just taken her PSATs and done well on them. Mark said his son was starting to look at colleges. "That's great," Bernie said.

Brown approached Bernie's apartment building on Lexington Avenue but couldn't pull up because a garbage truck was blocking the entrance. The three men got out of the car and walked a block to the grand prewar high-rise. They entered under its stately green awning, smiled politely at the doorman, and rode the elevator up to the Madoffs' penthouse. Walking on the huge gold and red Lavar Kerman Persian carpet, past a room filled with antique furniture and wall fixtures, they settled into chairs as Ruth hovered nearby.

Bernie faced his two sons. "I'm finished," he said. "I have absolutely nothing." His fabled investment business was "basically, a giant Ponzi scheme."

"It's all just one big lie," he added.

Mark was 44 and Andy was 42. Their father had lied to them about the family business their entire lives. His renown as a genius investor, the mystique that made them both venerate and fear him, was based on a fabrication. He was a fraud.

As Andy sat on the floor, sobbing, Bernie spelled out the implications. The company was insolvent, and had been for years. He'd been propping it up with stolen money. Some $50 billion in investor assets were gone. Most of it had never existed to begin with.

Virtually every friend and relative of the Madoffs had invested in Bernie's fund, including Bernie's sister, Sondra, and Ruth's sister, Joan. Countless employees were invested. The number of ruined lives was too high to comprehend.

Bernie asked for one more week to distribute what was left of the company's money to certain investors. Then he would turn himself in. Even now, he didn't seem to recognize that it was other people's money he was spending.

Forty-five minutes after he dropped them off, Clive Brown saw Andy and Mark emerge from the building. "We're not going to be riding back with you," Mark told him. "We're going to have lunch. Dad will be down soon."

At some point in the day, Bernie called Ira Sorkin and postponed their meeting, which was supposed to take place two days later. Madoff set it for the following week, presumably after he'd written his checks and was ready to turn himself in.

Bernie came downstairs half an hour after his sons left. On the ride back to the office, he got on his cell phone; Brown didn't know who was on the other line. "Andy's scared shitless," Madoff said. "I'll see you shortly." When they arrived back at the Lipstick Building, Bernie seemed his old self. "Thanks very much, Clive," he said. "Enjoy the rest of the day."

• • •

The Christmas party began at 6 p.m. It was held at Rosa Mexicano, a grand, upscale Mexican restaurant on First Avenue and Fifty-eighth Street, walking distance from the Lipstick Building. The company took over the entire establishment for its employees on this Wednesday evening, and paid for a good party. There was a taco station, a guacamole station, a buffet bar, and waiters walking around with frozen pomegranate margaritas, two of which could put a person out for the night.

Madoff's traders, secretaries, computer technicians, clerks, and managers soon filled up the large dining room and milled around its bright red and pastel-colored furniture and traditional Mexican artwork on the walls.

The family filed in. Bernie arrived with Ruth, both smiling as if everything were perfectly fine. She wore a spectacular black ensemble that made Elaine Solomon gush with envy. It was characteristically understated, though any woman could tell it was enormously expensive. "God," Solomon thought, "I wish I could wear her clothes." Peter and his wife, Marion, arrived, and then Shana.

Bernie and Ruth spent much of the evening sitting at the corner of the bar, trying to keep away from the crowd. At one point, they decided to sit down with the drivers, which was akin to joining the help. They were soon joined at the table by Peter, Marion, and Shana. "Boy, we're sitting at the royal table tonight!" exclaimed Errol Sibbley, Mark and Andy's driver.

Predictably, a line of employees discovered the new power

table and migrated toward it, trying to steal some face time with the boss. "Look at these jokers," Bernie laughed to the drivers. "I'm trying to come over here and talk to you."

There seemed to be a deep and unusual amount of emotion swirling through the family this evening. "There was a lot of hugging," Sibbley said. The company had hired a tarot card reader, and she sat down with Shana and started to read her cards. "You won't have to worry about money for the rest of your life," she said.

Only Andy and Mark were missing, a topic of whispers among the traders milling around the bar. "Where are they?" Solomon asked Sibbley.

"I don't know."

"You have to know. You're their driver."

All he could tell her was that a few hours earlier, he had picked up Mark at his Mercer Street loft in SoHo and brought him to a building on East Forty-ninth Street, where he was met on the street by Andy, who looked anxious, "like he was seeing a terrible accident or something," Sibbley said. The brothers ducked into a building, and Sibbley waited downstairs for ninety minutes. Finally, Mark called his cell phone and told him to take off. "Go to the party," he said.

It's not clear whom the brothers went to see, but at some point after their traumatic meeting with their father in the morning they decided they needed to speak with a lawyer. They called attorney Martin Flumenbaum, who rushed back to the city from a court appearance in Hartford, Connecticut to meet with them. When they spoke, he told Mark and

Andy they had no choice but to contact the feds and turn their father in. Bernie had implored them not to do that, but his wishes meant little to them anymore. Flumenbaum called the FBI, the SEC and the U.S. Attorney's office. An SEC attorney did a double-take when Flumenbaum described the case. "Did you say millions or billions?" he said.

At 8:30 the next morning, December 11, FBI Special Agent Theodore Cacioppi and an associate arrived at 133 East Sixty-fourth Street. They walked into the elegant, understated lobby of Madoff's apartment building, past some leather chairs and a single orchid, and introduced themselves to the building's doorman. He escorted them to the elevator to the penthouse. Bernie Madoff was in his pale blue bathrobe and slippers when Cacioppi knocked on his door and asked to come in.

A six-year veteran of the Bureau, Cacioppi told Madoff that he was aware of his confession to his sons the previous morning. "We are here to find out if there is an innocent explanation," the agent said.

Madoff stared at him with his small, expressionless eyes and took a breath. "There is no innocent explanation," he said.

There was no need for Cacioppi to pry information out of him; it flowed freely. He had "paid investors with money that wasn't there," Madoff said. He was broke. He expected to go to jail.

About an hour later, as Andy and Mark sat at an attorney's office in Manhattan giving sworn depositions to government lawyers, Madoff's attorney Ira Sorkin was sitting in a toddler's chair in a nursery school in Kensington, Maryland, watching his granddaughter play with her classmates. He wore a finely tailored suit and big round glasses under a head of moppy, snow-white hair; with his old-school formality, he was a Hollywood-worthy vision of a grizzled defense attorney.

"Let's all gather around and discuss a farm," the teacher announced to her class of toddlers. "What sound does a cow make?" she asked, holding up a picture.

The classroom answered in unison. "Moooo!"

"What about a duck?"

"Quack!"

Sorkin's cell phone rang. With the children in the background still making animal noises, he answered the call.

"Ike, it's Bernie Madoff. I'm handcuffed to a chair. I've been arrested by the FBI. I need your help." He was at FBI headquarters in Lower Manhattan, awaiting arraignment.

Attempting to digest what he was hearing, Sorkin told Madoff to put him on the line with the FBI agent sitting with him. The lawyer cut off questioning until he could catch a plane back to New York. But it was too late. Madoff had confessed.

At about the same time, Elaine Solomon arrived at work, later than usual due to the previous night's party. Eleanor Squillari had been there for several hours already, and met her with a grim expression. "Elaine, the FBI and the SEC are downstairs," she said.

"Why?"

"I don't know."

Then Peter walked through the glass doors of the executive offices with his briefcase. He was smiling, almost beaming, as he hung his coat in his closet. Solomon had rarely seen him so happy in the three years since Roger's death. "He's in such a good mood today," Solomon whispered to Squillari.

Squillari approached him. "The FBI and the SEC are downstairs," she said quietly.

"Okay," he said, without registering emotion. He walked toward the spiral stairs leading to the eighteenth-floor reception area.

The secretaries threw out some guesses about what the feds were doing there. "It must be a tax thing," Solomon said, "or something with compliance." And then the phone rang, and a colleague of Solomon's told her that Bernie was in custody.

Peter walked into the trading room and asked for quiet. The room came to a standstill. "Bernie's been arrested," he told the traders gathered around him. "I don't know why. I have no other information." And then he walked away, entered his office, and shut the door.

In the winter of 2003, the company's drivers were so busy ferrying various cancer-stricken Madoff family members to their oncologist appointments that they joked they were more like EMS drivers. With the company expanding at the time, Peter decided it needed another chauffeur.

Errol Sibbley was a well-dressed, well-mannered, middle-aged Jamaican immigrant. He'd been working as a Volvo mechanic in White Plains when his brother, who was already working for the company driving Bernie around, told him about an opening, driving for Mark and Andy. Since then, Sibbley had come to like the boys; Mark was friendly, and even Andy had his charms despite his wooden demeanor.

A few hours after Bernie was led out of his apartment building in handcuffs, Mark called Sibbley's cell phone and told him to drive down to his apartment in SoHo. His building was not for Manhattanites on a budget. Jon Bon Jovi occupied a duplex penthouse he'd purchased for $26 million. David Geffen had an ownership stake in a $10 million loft that Calvin Klein's daughter Marci occupied. Vera Wang occupied a commercial space.

Sibbley arrived at Mark's loft ahead of his boss. He was let in by Mark's wife, Stephanie, holding a child in her arms. She led Sibbley into their living room, where he was met by the couple's personal household staff, a housekeeper and nanny. "Make yourself comfortable," Stephanie said. She ordered Chinese food for them.

Mark arrived looking as though he'd been crying. He thanked the three employees for joining them and plunked himself down on a sofa. He took a breath and tried to compose himself. "I'm not going to be able to keep you anymore," he told them.

He started to sob. With each sentence, he stopped speaking and attempted to pull himself together. "All along, what Dad was doing was a big lie. There was big money involved. He may go to jail. . . . I'm afraid we won't be able to live this lifestyle anymore. I can't believe what's happened."

"There'll be maybe one more paycheck for you," Mark told Sibbley. "The government is going in there. They've taken over the office."

The sons were financially destroyed. Mark and Andy had $40 million each in company delayed compensation accounts that would soon evaporate as quickly as the life savings of Madoff's clients. Their father had issued them loans for tens of millions of dollars that he planned to forgive; they would now be listed instead as company assets to be re-paid. Andy had over $7 million invested in Madoff's investment advisory fund that vanished with the Ponzi scheme's collapse. The company the two helped run was once worth $1 billion; now it was going down the drain.

His employees stared at Mark in disbelief. "There's no more Madoff?" Sibbley asked.

"No."

Stephanie was furious. "I respected this man," she said of her father-in-law. "I worshipped him. I looked upon him as a father. And all along it was a big lie. Did he think about the grandkids when he did these things? He's fucked up our lives!"

A sense of unreality hung in the air. With not much left to say, everyone rose from their seats to leave. There were tears and hugs as Mark, Stephanie, and the people who had made life so comfortable for them bade farewell to one another. They put on their coats and streamed out. When the door behind them closed, Bernie Madoff's son and his wife found themselves alone.

It wasn't until later in the day that the news went public. Up at the office, someone raced into the executive suite after the closing bell and told his colleagues to head to the television monitors.

"If you work on a trading desk, stop what you're doing for one second before you walk out the door and clean your desk out for the day," said CNBC's Michelle Caruso-Cabrera. "Bernie Madoff has been arrested." The words *giant Ponzi scheme* flashed at the bottom of the screen. The criminal complaint against Madoff stated that he'd been turned in by "two senior executives" at the company, she reported.

The employees were in shock.

"It's a mistake, an absolute mistake," Solomon said.

And then something occurred to all of them: Andy and Mark hadn't been to the office all day. They'd missed the Christmas party the previous night. They had turned in their father.

Frank DiPascali was among those watching. He'd been pacing nervously through offices all day, but when he saw Madoff's image on television he turned pale. It wouldn't be long before the FBI turned its attention to him. He backed away from his colleagues, rushed into Bernie's private bathroom, and threw up.

The rest of the office staff remained at the television, taking in the bulletins scrawling across the screen. Photos and video of their boss looped repeatedly. Bernie was gone, vaporized into history.

11

Shattered Lives

On the morning following Madoff's arrest, it fell to the secretaries to contain the panic. To even get to the office, Eleanor Squillari and Elaine Solomon had to push their way through a mob of angry clients, reporters, and camera crews crowding the lobby. Once upstairs, they encountered the din of eight phone lines ringing at once. Fax machines were spitting out piles of redemption orders. The hallways were swarming with federal agents.

Virtually all the calls coming in were from desperate investors demanding their money. Some raised their voices. Others broke into tears. Solomon cut off each caller before the conversation got too involved. "I'm so sorry," she told them in her sympathetic British accent. "Let me just send you downstairs." But no one on the seventeenth floor picked up the phone.

Squillari tended to get sucked into their stories. "Eleanor, the phones are ringing!" Solomon yelled. Trapped into listening to the desperate stories, Squillari could only shrug.

Eventually Squillari could no longer cope with the endless calls, and stormed downstairs to the seventeenth floor to demand that the staff answer the phone. When she arrived, she told *Vanity Fair*, the place was deserted but for Frank DiPascali, sitting in his jeans and topsiders, talking on his cell phone.

"Frank, the phones won't stop ringing!" Squillari said. "What should I say to them?"

He looked up at her. "Tell them nobody's available," he said.

The anarchy was so consuming that Solomon could barely focus on the personal implications of Madoff's arrest: she'd invested her 85-year-old mother's life savings with him.

Finally, Peter walked in. "Where have you been?" Solomon asked.

"I had to go to my brother's and sign the bond for $10 million," he said.

"You did what?" she snapped. "You cosigned his bond?"

"Elaine, don't go there. I've had enough with my wife."

Squillari begged him to take control of the chaos. "What am I supposed to do with these people that want redemptions?" she asked.

"I don't know," he shrugged. No one wanted to tell them their money was gone.

Tensions continued to rise in the following days as investigators trampled over an office that Squillari and Solomon had kept meticulously organized. The secretaries watched as investigators spilled drinks on furniture and threw papers around without apology.

Solomon recognized an SEC staffer who had visited the office a few years earlier, in one of the agency's brief inquiries into Madoff. The mild-mannered Brit lost her temper. "If you had done your fucking job in the first place we wouldn't be in this position now!" she yelled.

It all ended for Solomon when she lost her temper one too many times. One of the investigators knocked a cup of coffee to the floor and walked away without cleaning it up. "Would you leave your house like this?" she demanded. The next day, the feds let her go. "You're done," she was told.

It was Christmas Eve. Solomon said goodbye to the few remaining colleagues there, packed her things, and headed downstairs, leaving the Lipstick Building behind. Her career at Madoff Investment Securities had ended with FBI agents searching through garbage cans and her boss under house arrest. Thanks to Bernie Madoff, she was unemployed and her mother was destitute. But as she walked down Third Avenue, pushing through crowds of people rushing home with their Christmas presents, Solomon found it hard to work up any anger toward him. "I couldn't hate Bernie," she recalled months later. "I don't believe that he set out to do it intentionally. He and Ruth were such nice people."

• • •

The day of the arrest, Carl Shapiro was in his living room in Palm Beach when Robert Jaffe called. "Turn on CNN," he said.

The old man flipped on the television. Bernie's face flickered on the screen, along with reports about his Ponzi scheme. Shapiro was stunned. He'd seen him a few weeks earlier on Thanksgiving weekend. He had just loaned him hundreds of millions of dollars.

The implications were incomprehensible. Between his personal investment and his foundation's, Shapiro was out a half-billion dollars. Dozens of his friends and acquaintances were wiped out. A legion of charities lost their endowments. Jaffe, Madoff's Palm Beach operator, was disgraced. Bernie had made fools of them all.

The moment felt like "a knife in the heart," Shapiro said. He was 95 years old; friends worried that he would have a coronary.

"How could I have been so wrong for fifty years?" he'd ask people in the weeks that followed. "How could we have been so stupid?"

The wreckage was epic. Madoff and Jaffe had worked the Palm Beach Country Club so thoroughly that a third of its members invested with them. The losses were in the billions of dollars.

Palm Beach accountant Richard Rampell witnessed the carnage. Unable to talk some of his clients out of investing with Madoff, he watched them stagger into his office shell-shocked after incurring losses ranging from $2 million to $80 million.

Madoff's best friends were horrified. After decades going out to dinners and movies with Bernie and Ruth, one of his closest confidants couldn't comprehend that Bernie had been sending him fabricated financial statements all along. So complete was his trust in Madoff that he had invested tens of millions of dollars in an investment formula he didn't even understand. "Bernie would never give me details that I could comprehend," he said. "He would say 'I don't even know how to explain it. The system is doing it.'" Despite all the weddings and bar mitzvahs celebrated together, the vacations spent with their wives and the lazy weekends sailing on the *Bull* together, he was left with almost nothing, as were his family members. "He had all my children's money, he had all my grandchildren's money," he said. "Every nickel of it was there in the account. We just gave him everything." The betrayal was too horrible for him to even acknowledge. "I wouldn't be shocked if Bernie walked in here now and said 'How do you like this bullshit—it's all a mistake,'" he said. "I'd say, 'I always knew it.'"

The atmosphere in Palm Beach turned funereal as the victim count climbed into the dozens and then into the hundreds. Members of the Palm Beach Country Club walked around in a daze, conned by a man they had treated as a deity.

Their personal losses were transcended only by the catastrophe befalling the charities they championed. Nothing since the Great Depression equaled the devastation Madoff visited upon America's Jewish nonprofits. Yeshiva University, led into investing by Ezra Merkin, watched over the years as its $14 million investment with Madoff grew to $110 million—and then vanished. *New York Daily News* publisher Mortimer Zuckerman's Charitable Remainder Trust lost $25 million in a Merkin feeder fund; he claimed another $15 million personal loss. The Robert I. Lappin Charitable Foundation closed its doors after losing $8 million. The Ramaz School in New York lost $6 million.

Prominent Jewish benefactors such as the director Steven Spielberg and his partner, Jeffrey Katzenberg, saw their foundations financially devastated. Jewish philanthropists in Minnesota, drawn to Madoff by Mike Engler, lost fortunes. Merkin's investors, many if not most of them prominent New York Jews, lost $2.4 billion in Madoff's fraud. Almost half of that came from the accounts of Merkin's fellow Fifth Avenue Synagogue members.

On December 11, Elie Wiesel discovered that the $15.2 million invested by his Elie Wiesel Foundation for Humanity had been wiped out. Forty years' worth of personal income from his books, lectures, and university salaries disappeared as well. "We thought he was God," Wiesel said. "We trusted everything in his hands."

Ponzi schemers characteristically search for "affinity groups" to spread word of their investment magic through the grapevine. Madoff had used the Jewish community as an incubator for his scam. As news of the Madoff scandal grew, the Internet swelled with anti-Semitic vitriol. "It is not hateful to notice that the majority of people guilty of this in all the stories coming out are Jews," a reader commented on a *Palm Beach Post* reporter's blog. "A group that at most makes up only 3% of the population seems to have wormed their way into our financial system and are bringing it down." Madoff had not just bankrupted his investors; he had sold out his own people.

The seething anger among his Palm Beach victims was bound to come out in some form, and two days after his arrest, a 78-year-old retiree led the charge.

At the height of his success, Jerome Fisher's Nine West brand sold one out of every five pairs of shoes in America. He was a Palm Beach fixture and a benefactor to a slew of Jewish charities and universities. When Madoff confessed, $150 million of Fisher's fortune went down the drain.

Two days after the arrest, a subdued but dutiful crowd filed in for carpet magnate John Stark's black-tie birthday celebration at Donald Trump's Mar-a-Lago estate. Many guests in the crowd had lost millions of dollars in the debacle. Fisher's eyes widened when Robert Jaffe made an entrance, debonair as usual in his crisp tuxedo and slicked-back hair. Fisher grew enraged, for it was Jaffe who had greased the wheels for his investment in Madoff. "You've got a lot of nerve showing up here!" Fisher yelled, according to some accounts, bringing the party to a standstill. Jaffe was stunned. "You got a fucking commission—a point and a half—on me!" Fisher reportedly shouted. People stepped in to quell the scene as others, including Donald Trump, looked on. Once one of the most sought-after dinner guests in Palm Beach, Jaffe had become a pariah.

In a drab asphalt West Palm Beach strip mall called Western Plaza, business started picking up at the storefront headquarters of Royal Pawn and Jewelry. A parade of well-dressed women from the tasteful side of the Intracoastal Waterway made their way past drum sets, exercise bikes, used DVDs, and old power drills to a dusty glass counter, where they plopped down fistfuls of diamond rings and jeweled necklaces. The riches had been acquired in happier times, gifts from husbands or boyfriends back when the men brimmed with confidence, pride, and money.

A society matron came in to sell off every piece of artwork on her walls, every piece of jewelry in her dresser drawers, every vestige of status and wealth she had accumulated over half a century. An auction house would have fetched a far better price for it all, but she was almost broke, with little time to waste. She was left to haggle with a manager to get fifteen cents on the dollar.

For a lot of people, the image was of a group of stratospherically rich people who had lost money in their greed for even more. "Someone who has tens or hundreds of millions and who is at the end of their life and is still trying to get as much as they can is just disgusting," wrote one blog contributor, apparently referring to Carl Shapiro.

But the pain in the real world was, if anything, more acute.

When the phone rang at Stephen Richards's house in Boca Raton, it was his nephew in Manhattan. "Uncle Steve, Bernie Madoff was arrested."

"For what?" Richards asked. "DWI?"

"No, it's for fraud."

Stephen and his wife, Fran, turned on the television, probably at the same time as Donny Rosenzweig, Cynthia Arenson, and thousands of others who had placed their future in Bernie Madoff's hands. The Richardses watched the news with anxiety digging a pit in their stomachs. Almost every cent they had was in their Madoff account. The moment the anchor mentioned the words *Ponzi scheme*, Stephen realized their lives had changed. He looked at Fran. "We have nothing," he said.

"I don't believe this!" she said. She bolted to the telephone and called Joan Roman, Ruth Madoff's sister. It had been fifty-five years since Stephen had introduced Joan to his Ohio State fraternity brother, Bob Roman, on that porch in Laurelton, a meeting that until this phone call had been perhaps the luckiest day in all of their lives. "Did you see Bernie was arrested?" Fran asked her.

"Get out of here!" Roman replied. Ruth hadn't called her.

"I'm telling you, put on the TV."

Joan and Bob turned on their television, according to Fran, and saw recurring images of Joan's brother-in-law on the screen. The couple was "hysterical," Fran said. The idea that Bernie would have used his sister-in-law's money to help fuel a Ponzi scheme was unfathomable—except that he had done the same thing to at least one of his own sons. Just a few miles away from the Romans, Bernie's sister, Sondra, was discovering that he'd swindled her as well.

Stephen Richards lost $5 million. The Romans lost $11 million. Part of the numbers represented profits that were a figment of Bernie Madoff's imagination. But Madoff paid out redemptions from those profits for years, and Richards and Roman counted on them to retire, buy homes, and support their families. As far as they knew, the money in their Madoff accounts was as real as their checking account balances.

To hold the futures of his family members and best friends in his hands must have been a heady feeling for Bernie; he was taking care of all of them into their old age. But he'd done it by dangling their lives over the precipice of disaster. Perhaps he enjoyed the risk of it all, as he did when he sped his car down an icy road with a terrified young friend beside him.

Madoff had given and taken from them, as though he were God. Even after things collapsed, he attempted to choose whose suffering he would ease. Soon after investigators moved into his office on December 11, they discovered $173 million worth of uncashed checks inside his desk drawer, made out to friends, employees, and family members. It was apparently part of his plan, as he'd outlined it to Peter, to distribute money to certain people before he turned himself in. Andy and Mark cut those plans short by turning him in.

Sitting in his penthouse apartment under house arrest in December, he made a last effort to give some financial relief to people of his choosing. A few weeks after his confession, he and Ruth sent out five care packages to relatives, including Ruth's sister, Joan, and Bernie's sister, Sondra. One package contained thirteen watches, a diamond necklace, an emerald ring, and two sets of cuff links. The other two contained a diamond bracelet, a gold watch, a diamond Cartier watch, a diamond Tiffany watch, four diamond brooches, a jade necklace, and other jewelry, according to the government. Madoff's bail was almost revoked for violating his pledge not to disperse any of his assets, and his relatives had to return the gifts. His last-minute attempts at charity accomplished nothing for his traumatized family members and even less for the thousands of other victims for whom he showed no mercy.

Stephen Richards was 75 years old, broke, and on kidney dialysis. Both he and Roman had led good lives thanks to their chance encounter in 1953 with a young Bernie Madoff. Half a century later they were senior citizens, chauffeuring people to the airport for a few bucks in order to feed themselves.

Harry Markopolos was at his twin sons' karate class in Massachusetts on the evening of December 11. Amid the grunts and thuds, he glanced down at his phone and saw that he had two voicemail messages. He ducked outside.

Both messages were from friends calling with the news of the arrest. Nine years after Markopolos had concluded that Bernie Madoff was a crook, his nemesis had finally confessed. Markopolos felt a "huge burst of energy" and then almost fainted. He grabbed a railing in order to keep himself standing.

Markopolos was never able to expose Bernie Madoff—the mastermind simply gave up when his scheme collapsed—but the cancer in the financial system that Markopolos warned of for so long had finally been removed. Yet a sense of imminent danger never loosened its grip on Harry Markopolos for very long. As his rush of joy abated, he quickly became fearful, this time of the SEC. "I was very, very worried that the SEC was going to come in and seize my computers and my documents because they knew what I had turned into them and perhaps they wouldn't want that made public," he recalled. "And so I told my family I was worried for their safety." He raced to get his documents published "because I didn't want my home invaded by federal agents taking the documents out and destroying the evidence." He called the *Wall Street Journal*.

On December 18, 2008, America discovered Harry Markopolos. The *Journal* put a sketch of his face on its front page and told his story. It was a riveting description of the ups and downs of his odyssey to get the government to listen to him, peppered with dramatic quotations. "I kept firing bigger and bigger bullets but I couldn't stop him," he said of Madoff. A hero was born. *60 Minutes* profiled him; movie offers streamed in. Markopolos testified at congressional hearings into the Madoff affair, at which a congressman pressed him to consider heading up a branch of the SEC. He declined.

Perhaps most rewarding of all for him, SEC Chairman

Christopher Cox basically admitted that the agency had been wrong to doubt him. "The Commission has learned that credible and specific allegations regarding Mr. Madoff's financial wrongdoing, going back to at least 1999, were repeatedly brought to the attention of SEC staff, but were never recommended to the Commission for action," Cox stated. "I am gravely concerned by the apparent multiple failures over at least a decade."

That was an understatement. The SEC's Inspector General's office found that the agency received a stream of tips and letters over the years alleging that Madoff was running a Ponzi scheme, and not just from Harry Markopolos. "If my suspicions are true, then they are running a highly sophisticated scheme on a massive scale," wrote one informant. "Your attention is directed to a scandal of major proportions," wrote another.

In each instance, the agency either ignored the warnings altogether or reacted to them with slapstick incompetence, missing some clues that were right before their eyes while ignoring others out of sheer laziness. Elaine Solomon was right: the SEC staffers were untrained, fresh out of school and in total awe of Bernie Madoff, according to the Inspector General's report.

No one was more shocked at the agency's ineptitude than Madoff. In 2006, SEC investigators asked him for his account number at the central Wall Street clearinghouse for stock transactions. It would have taken a single phone call to learn that Madoff had been fabricating all his trade results. "I thought it was the end, game over," he recalled in a jailhouse interview with the Inspector General's staff. But no call was placed. Madoff was "astonished" at their blundering.

Vindicated at last by the SEC's findings, Harry Markopolos plunged into his new career as a private fraud investigator—with "a very private office with heavy security," he said.

On his European trip, Markopolos had worried that private investors and banks would be devastated by Madoff, and his prediction came true.

Walter Noel learned about Madoff's arrest at his Manhattan office on December 11. After watching the television news bulletins with horror, he tried desperately to reach Madoff by phone but was unsuccessful. He then called his wife and suggested she sit down before he told her some news.

Fairfield Greenwich's accounts with Madoff were worth almost $8 billion, all of it washed away in the blink of an eye. Much of Noel's own worth went with it. Fairfield Greenwich—or, more accurately, its investors—became the single biggest loser in the Madoff scandal. All the princes and princesses, dukes and duchesses romanced by Noel's glamorous, globetrotting family were fleeced by a middling intellect from Queens. Banco Santander's investors lost $3 billion; Fortis Bank in the Netherlands lost $1.4 billion; Union Bancaire Privée lost $700 million. Sonja Kohn's Bank Medici lost $2.1 billion and was taken over by the Austrian government.

But no one lost as much as René-Thierry Magon de la Villehuchet.

On December 9, at about the time that Madoff was revealing his secret to his brother, an old friend ran into de la Villehuchet at the Jockey Club in Manhattan and started to wring his hands about the economic crisis. The elegant, silver-haired aristocrat lightheartedly brushed off the question: "Yes, it's terrible, but thankfully for us, we are with Madoff."

Two days later, de la Villehuchet was watching television when he learned that the FBI had arrested Madoff. Alarmed, he called his partner, Patrick Littaye, who at the time was in Saint-Malo, on the rugged northern coast of Brittany. The two were stunned. Then de la Villehuchet went into survival mode, huddling with his band of European royals at Access International and consulting lawyers and other financial specialists, hoping to recoup the staggering losses for his clients and himself. He felt deeply betrayed by Madoff, and ashamed that he had betrayed his clients.

By December 18, de la Villehuchet realized that he had neither the means nor the time to rebuild his firm. His twenty-eight employees could not be paid, and the rent was due. There were rumblings of possible prosecution in Europe. "You're toast," a friend told him. "Your professional life is over."

Guy Gurney, a British photographer and a sailing friend of de la Villehuchet's, was scheduled to have dinner with him shortly before Christmas to talk about an upcoming regatta.

When he called to change the date, de la Villehuchet was choking back tears. "I cannot speak," he said. "I'm involved with Madoff."

"He sounded devastated," Gurney said. Madoff's collapse had destroyed de la Villehuchet's firm and taken $50 million dollars of his personal fortune with it.

On December 22, 2008, less than two weeks after the public learned about Madoff's Ponzi scheme, he called his wife, Claudine, from his Madison Avenue office to say that he had a late dinner engagement and not to wait up for him. He asked the janitors to finish cleaning the office by 7 p.m. Then he locked his door.

The next morning, Claudine called a colleague of her husband's, worried that he hadn't returned home. When the colleague arrived at the offices of Access International, Thierry de la Villehuchet lay in his swivel chair, one leg on his desk, the other on the floor, his wrists and arms slashed with a box cutter, a bottle of tranquilizers open on the desk. The dapper 65-year-old banker had carefully placed a trash can to collect his spilling blood.

He left no suicide note, just a letter to his wife, insisting on a private cremation, hoping against all odds for a discreet exit.

His would not be the only suicide. Two months later, William Foxton, a decorated former British solider, sat on a park bench and shot himself in the head after losing his life savings to Madoff.

• • •

By January 2009, Bernie Madoff was the most hated man in America. The public was outraged that he had been allowed to stay at his Lexington Avenue penthouse while out on bail. His transformation from Wall Street royalty to notorious villain was cemented six days after his arrest. Met by a voracious press pack as he returned to his apartment from a court date, he wore a baseball cap, a winter coat with an upturned collar, and a disturbingly oblivious smile. He tried to push his way through the media horde, only to have an angry photographer forcefully push him back. It was a televised moment of humiliation for a man who was once so powerful that people were barely allowed to see him, much less touch him.

Madoff received so many death threats that he started wearing a bulletproof vest to his court appearances. Protestors sometimes gathered outside his apartment building, including an angry Wall Street trader who arrived on January 14 with a huge sign reading, "Bernie, it's not too late to do the right thing: JUMP!" The *New York Post* put the man and his sign on the front page the next day.

The Madoff family was ripped apart. After telling friends they were so furious at their father's deception that he was dead to them, Andy and Mark went into hiding. Ruth became a pariah. Before Bernie went to jail in March, he made a desperate plea for her to be able to keep her possessions, the penthouse apartment, and $70 million in her bank accounts. She routinely visited her husband in jail and took months to express remorse for the victims rendered penniless by him. As a result, she became a despised woman, banned from her favorite restaurants and her hair salon, hounded by paparazzi, and jeered in the tabloids. The *New York Times* dubbed her "The Loneliest Woman in New York." She eventually lost all her homes, including the apartment, and agreed to forfeit all but $2.5 million in savings, which would in all likelihood be targeted by a long line of claimants.

There was not much question that Bernie would plead guilty—he had already confessed to the crime—and on March 12, 2009, he showed up at the federal courthouse in Lower Manhattan, walking past a satellite village filled with reporters and camera crews from around the world. As he stood before Judge Denny Chin and admitted his guilt—"I knew what I was doing was wrong, indeed criminal," he said—a sad parallel was taking place in Florida.

When Neil and Constance Friedman in Palm City learned the horrifying news about Madoff, the first call they made was to Jerry Horowitz. The Friedmans owed their prosperity to their old friend, who, as Bernie Madoff's veteran accountant, was able to get them into Madoff's fund. Eighty years old and stricken with cancer once again, Horowitz had already heard the news from his son-in-law and Madoff's accountant, David Friehling, who himself was scrambling for information. "Maybe we'll be all right," Jerry said. "I don't know."

Sitting in their pajamas, the Friedmans hugged one another and cried. They lost $4 million—all of their money and most of their children's. Their comfortable retirement ended with one news bulletin. They were left to scour the papers for dollar meal specials at local fast-food restaurants.

After decades spent auditing Madoff's books by himself out of his home, Horowitz too was wiped out by Bernie Madoff. The months following the arrest became "a living nightmare" for him, according to his son. He was a dying old man thrown into the vortex of an international scandal.

The day Madoff admitted his guilt, Judge Chin revoked his bail and sent him to jail to await sentencing. It was a moment of at least minor satisfaction for many of those he'd betrayed. On that same day, Jerry Horowitz died of cancer, penniless and heartsick over his old friend's larceny. Whatever secrets he held about Madoff died with him as his old client was led away in handcuffs.

Madoff choreographed his downfall as carefully as his rise. He'd told investigators from the day of his arrest that he expected to go to jail, and he seemed determined not to drag anyone else with him. He pleaded guilty to all eleven charges against him; with no plea bargain, he was under no pressure to name names.

The government, however, was pressed to expose accomplices, and it offered up Madoff's enablers. They arrested Friehling, his auditor and Jerry Horowitz's son-in-law, for alleged securities fraud, though without alleging that he was in on the scheme (he pleaded guilty not long afterward). The SEC filed suit against Stanley Chais, Robert Jaffe, and Maurice "Sonny" Cohn, cofounder with Madoff of Cohmad Securities, claiming they turned a blind eye to Madoff's corrupt practices while making obscene profits off of him. Ezra

Merkin faced a similar suit by New York State Attorney General Andrew Cuomo.

Both the Massachusetts Secretary of the Commonwealth and court-appointed trustee Irving Picard sued Walter Noel and Fairfield Greenwich for allegedly turning a blind eye as well. Picard also sued an investor and friend of Bernie's named Jeffrey Picower, who may have made more money from the investment enterprise than anyone, including Madoff himself. The trustee charged that Picower withdrew $5 billion from his Madoff account over the years, flying in the face of his claim that he lost money. How much Picower knew about the scam may never be known; ten months after Madoff's arrest, Picower was found dead at the bottom of the swimming pool at his Palm Beach estate. An autopsy determined he'd suffered a massive heart attack.

All of Madoff's former feeders and salespeople claimed they had been lied to and deceived by him. All of them denied having any knowledge of the scam. Most said that they had lost millions to Madoff and were actually victims, not perpetrators. After Bank Medici was taken over by Austrian bank regulators, prosecutors said Sonja Kohn had received $40 million in kickbacks to funnel customers to Madoff. "I am actually the greatest Madoff victim," she responded. Ezra Merkin denied accusations that he had hidden Madoff's role in his investment funds. Fairfield Greenwich claimed it "conducted vigorous and robust monitoring on an ongoing basis of the Madoff investments," contrary to the claims of lawsuits against the company. "Like everybody else who trusted and invested with Bernie Madoff, he betrayed my trust," said Stanley Chais.

In August 2009, Frank DiPascali pleaded guilty to 10 felony counts, including conspiracy and tax evasion. Three months later, computer specialists Jerry O'Hara and George Perez pleaded guilty and admitted they had designed the software that facilitated the scam. But these were supporting characters in the Madoff drama; the public hungered for action against his family members. Ruth, Peter, Andy, Mark, and Shana were the major players in Bernie Madoff's business and personal life. He was the center of their universe, tightly controlling the flow of information to them. Yet they were all capable of figuring out the scam if they had looked for it. Why they didn't is a question that may dog them forever.

On June 29, 2009, shortly before sunrise, a line started snaking around the Manhattan federal courthouse. By the time the doors to the building were opened, a huge crowd had gathered. So many journalists, spectators, and victims of Bernie Madoff's scam showed up for his sentencing that an overflow room was required.

The courtroom was grand, with marble and cherry columns rising thirty feet, big bronze chandeliers, and tall windows overlooking the East River. An intimidating squad of plainclothes law enforcement officers lined the perimeter, on guard for attempts on Madoff's life.

When the defendant entered the courtroom, looking

haggard and pale from his three months in jail, the officers surrounded him until the doors to the chambers were
locked.

Madoff was wearing one of his charcoal gray suits, probably purchased for five figures at Kilgour in London. But it
seemed too large on him now, as if he'd grown smaller. All
the same, Judge Chin's decision to allow him to don the suit
instead of prison garb allowed him a measure of dignity his
victims didn't appreciate.

Madoff sat down at the defendant's table near the front
of the room and stared straight ahead. As the courtroom fell
silent, some of his victims rose one by one to a microphone,
each urging the judge to imprison Madoff forever.

A retired corrections officer named Dominic Ambrosino
stood with his wife, Ronnie Sue. They had used the returns
from their Madoff account to buy a vacation motor home and
travel the country. The adventure came to an end when the
money evaporated. Now they were struggling to find enough
money to pay for gas. "The fallout from having your entire
life savings drop right out from under your nose is truly like
nothing you can ever describe," Ambrosino said.

Tom Fitzmaurice, 63, walked to the microphone with his
wife, Marcia. In a clarion voice that enveloped the room, he
demanded the maximum sentence for "this evil lowlife." "He
cheated his victims out of their money so that he and his wife
Ruth and their two sons could live a life of luxury beyond
belief," he said. "I am financially ruined and will worry every
day about how I will take care of my wife. Where will we

be able to live? How will we pay our bills? How will we get medical insurance?"

An old man in the audience, another former investor, started to sob. His wife patted him on the back as Fitzmaurice read a statement from his wife, standing beside him:

> I cry every day when I see the look of pain and despair in my husband's eyes. I cry for the life we once had before that monster took it away.
>
> Our two sons and daughter-in-law have rallied with constant love and support. You, on the other hand, Mr. Madoff, have two sons that despise you. Your wife, rightfully so, has been vilified and shunned by her friends in the community. You have left your children a legacy of shame.
>
> I have a marriage made in Heaven. You have a marriage made in Hell, and that is where you, Mr. Madoff, are going to return. May God spare you no mercy.

The biblical overtones of his crime were brought home when Burt Ross, a former mayor of Fort Lee, New Jersey, who lost $5 million in the scam, stepped forward with his wife. A bulldog of a man with thinning white hair, he had once turned down a $500,000 bribe from mob-connected developers. But this morning he sobbed with each sentence.

"What Bernard L. Madoff did far transcends the loss of money," he said. "It involves his betrayal of the virtues people hold dearest—love, friendship, trust—all so he can eat at the finest restaurants, stay at the most luxurious resorts and travel on yachts and private jets. He had truly earned his reputation to be the most despised person in America today."

He cited Dante's *Divine Comedy*, in which the Italian poet branded fraud the worst of sins. "[Dante] placed the perpetrators of fraud in the lowest depths of Hell, even below those who had committed violent acts," Ross said. "And those who betrayed their benefactors were the worst sinners of all, so in the three mouths of Satan struggle Judas for betraying Jesus Christ and Brutus and Cassius for betraying Julius Caesar. . . . We urge Your Honor to commit Madoff to prison for the remainder of his natural life, and when he leaves this earth virtually unmourned, may Satan grow a fourth mouth, where Bernard L. Madoff deserves to spend the rest of eternity."

One victim said that after losing all her money to Madoff's scheme, she had been left to scavenge through dumpsters for food. Another said his family lost all the money set aside for the care of his mentally disabled twin brother.

It's a rare occurrence when the powerful are called to account by the powerless. Outside the federal courthouse, the clamor for the heads of financial institutions whose irresponsibility almost brought down the economy had gone nowhere. Madoff was a proxy for their sins.

There was no one to comfort Bernie Madoff as the words of condemnation washed over him. Not a single friend or family member showed up in court to support him. No one wrote letters to the judge urging leniency. A year after basking in glory at Carl Shapiro's birthday party, he was a reviled man, facing the abyss alone.

The prosecution outlined his crimes and argued that he deserved no mercy. In response, Madoff's lawyer, Ira Sorkin, argued, implausibly, that he deserved to spend twelve years in prison, one short of what he said was his likely remaining life span.

Then Judge Chin invited the defendant to speak. Madoff rose with his back still to the audience, so that only the judge could see his face. The back of his suit jacket was creased. With his careful and expensive haircuts a thing of the past, his curly gray hair fell randomly down the back of his head.

He spoke with a rasp. His usual bravado, and the slight hint of sarcasm that inflected it, were gone. There was only contrition. "I cannot offer you an excuse for my behavior," Madoff said.

> How do you excuse betraying thousands of investors who entrusted me with their life savings? How do you excuse deceiving two hundred employees who have spent most of their working life with me? How do you excuse lying to your brother and two sons who spent their whole adult life helping to build a successful and respectful business? How do you excuse lying and deceiving a wife who stood by you for fifty years, and still stands by you?

The explanation he offered for his crimes was compelling.

> I believed when I started this problem, this crime, that it would be something I would be able to work my way out of, but that became impossible. As hard as I tried, the deeper I dug myself into a hole. . . . I refused to accept the fact, could not accept the fact, that for once in my life I failed. I couldn't admit that failure, and that was a tragic mistake.

He was casting himself as a victim who had fallen prey to his pride, an overachiever who'd been afraid to concede a mistake. But his plea for sympathy turned the facts on their head. He was not an overachiever who'd failed at something; he was an underachiever who had succeeded by lying.

Bernie Madoff grew up feeling inferior and strove from an early age to prove himself worthy using the only yardstick that he knew: the amount of money he could make. He found his vehicle for success in an illegal but hugely profitable enterprise, and he seemed to easily wall himself off from any guilt over its consequences. His scam went on for so long that immorality became a way of life for him, until it probably didn't seem immoral to him at all.

To admit that would have been a true act of contrition. But as he faced his punishment, it may have been easier for him to believe he was simply a successful man who couldn't admit his one business failure.

> I live in a tormented state now, knowing of all the pain and suffering that I have created. I have left a legacy of shame, as some of my victims have pointed out, to my family and my grandchildren. That's something I will live with for the rest of my life.

His wife, he said, "cries herself to sleep every night," which he said was another source of torment for him.

It was a morally disorienting moment. Madoff seemed a man, not a monster or even a thief. Just a sad man.

> There is nothing I can do that will make anyone feel better for the pain and suffering I caused them. But I will live with this pain, with this torment, for the rest of my life.
>
> I apologize to my victims. I will turn and face you.

He swiveled around to the crowd and looked out upon his victims for the first time. There was no eye contact, and he seemed frightened to glance at them for more than a moment. But he gave people a chance to see what he had become since his fall. The skin beneath his eyes drooped, his face was gray, and his body seemed frail. He was extremely nervous. He seemed a broken man.

The sympathetic image Madoff evoked was quickly punctured by Judge Chin. "Objectively speaking, the fraud here was staggering," Chin began. "It spanned more than twenty years. . . . This is not just a matter of money. The breach of trust was massive. . . . Here, the message must be sent that Mr. Madoff's crimes were extraordinarily evil."

He told the story of a woman whose husband died and left behind his life savings in Madoff's fund. When she visited Madoff after his death, Bernie put his arm around her in a fatherly way and told her not to worry. "The money is safe with me," he said. She was so comforted that she deposited even more money into the account, including her 401(k) and her pension funds. It vanished with the collapse of the Ponzi scheme. "The victims put their trust in Mr. Madoff," Chin said.

He instructed the defendant to stand.

"It is the judgment of this court that the defendant, Bernard L. Madoff, shall be and hereby is sentenced to a term of imprisonment of one hundred fifty years."

The courtroom burst into applause, and some victims in the audience hugged one another.

As Judge Chin delivered his order banishing him from society for his execrable crimes, Madoff stood and stared at him with that impassive look of his. Even at this low moment, with his terrible acts exposed to the light of day, he didn't seem particularly evil. Standing there in his old, expensive suit, you could still see in him the discerning old wise man, reluctantly agreeing to take one more person's money. There was no hint of a monster about him. He had the look of a decent man.

Acknowledgments

This project is the culmination of an extraordinary team effort. For months, my two researchers and I spent long days in a small office learning every detail of the Madoff story. By the end, we were almost completing one another's sentences.

For some reason, Michael Valerio allowed me to talk him into taking off a semester of college to help me with this book, and I will be forever grateful for it. He has a brilliant mind, a level head and a big heart, and he proved—under punishing deadline pressure—to be a first-rate journalist.

Likewise, Steve Marmon, an investment banker and a veteran reporter, lent his wildly perceptive analysis, financial expertise, and investigative talents to the effort. He produced oceans of information and made sense of it all. I appreciate his help and friendship.

For some reason, both my husband Kyle Froman and I chose to take a swan dive off a career cliff this past year. The last twelve months have been one of the scariest, most exciting periods of our lives. We've spent a large amount of time supporting each other in this time, and my respect for him has grown throughout this adventure. I want to thank him for making life so exciting. I love him like crazy.

The journalists in my life make me proud of my profession, and none make me prouder than Adam Nagourney of the *New York Times* and Claire Brinberg of CNN. I thank them for their help with this book, and for their friendship—life would not be the same without them. Justin Blake is not a journalist, but knows more about the business than many who are. His advice and help were critical. My dear friends Paul Lombardi and Jeff Soref are two of the smartest people I know, and I was so lucky to have them when I needed their support and advice. And thanks so much to my good and wise friend Ben Kushner.

Karen and Walter Boss displayed a hundred acts of kindness during this project. They couldn't have been more supportive or loving—as was Karen's wonderful father Irv Freedman—and I thank them from the bottom of my heart. I wrote much of this book in the dead of winter in my house in Fire Island Pines, and Jon Wilner and Bob Howard were there with support, inspiration, and friendship—I value it deeply. The amazing Lou Del Vecchio and his talented assistant Daniel Thorstad ran the Madison Fire Island Pines so well that I was able to concentrate on this book, and I love and respect them both for it. Dan Cochran, a cherished friend, shared his considerable wisdom for this book.

A handful of people were so generous they actually housed me and my researchers, and we are extremely grateful to them. Gina Quattrochi of New York's Bailey House, a wonderful woman, came out of the blue to donate office space and I thank her. My good friend Andrew Miller handed me the keys to his home in the early days of the project. Phil Friedman and his partner Manny Bulaclac housed and cared for me during my stay in Palm Beach. They are all generous people with big hearts.

Many of Bernie Madoff's victims opened their homes to me and spent hours sharing their stories so that the public would have a better understanding of the pain he caused. I particularly want to thank Dominic and Ronnie Sue Ambrosino, Harlene Horowitz, Stephen and Fran Richards, and Neil and Constance Friedman for their help. I'd also like to thank Elaine Solomon for sharing her incredible story with me. There are a legion of others who assisted me but who requested anonymity. You know who you are and I thank you.

My former colleagues Andrew Friedman, David Friend, Dianne Doctor, and Jim Rosenfield were enormously helpful, and I thank them.

I'd like to thank Jeff Farber, Nathalie Mattheiem, Lisa Reyes, and Peter Kos for their terrific work on this book. I was also helped by Roxanne Donovan and Nick Nicholson, who went out of their way to educate me on the subject. I'd also like to thank Jack Stephenson, the Kirtzman and Froman families, Dan Goldberg, Ben Jobes, Mark Halperin, Karen Avrich, David Chalian, Patrick Healy, Seth Weissman, Erika Roberson, Nancy Gubman, and Mitchell Newmark for their support and advice throughout.

I finally would like to thank the larger-than-life figures at HarperCollins, especially Jonathan Burnham, who trusted me with this important project, and Claire Wachtel, my loving, brilliant editor, who was with me every step of the way, along with her enormously competent assistant Julia Novitch, and did magnificent work. Beth Silfin, HarperCollins's gifted counsel, was also enormously helpful.

I'd finally like to thank my agent, Flip Brophy, who made it all happen again.

—*Andrew Kirtzman*

Chronology

1878 Bernie Madoff's maternal grandmother, Gussie Muntner, is born in Austria.

1882 Bernie Madoff's paternal grandmother, Rose Krantz Madoff, is born in Poland.

Bernie Madoff's maternal grandfather, Benjamin Munter, is born in Poland.

December 12, 1883 Bernie Madoff's paternal grandfather, David Solomon Madoff, is born in Poland.

1900 Benjamin Munter emigrates to the United States from Austria.

1905 Gussie Munter emigrates to the United States from Austria.

April 28, 1907 David and Rose Madoff sail for the United States on the *Graf Waldersee* from Hamburg, Germany.

May 11, 1907 David and Rose Madoff reach Ellis Island. Naturalization papers list them as soon settling at 631 Costello Court in Scranton, Pennsylvania.

June 8, 1910 Bernie Madoff's father, Ralph, christened Zookan, is born in Scranton.

December 25, 1911 Bernie Madoff's mother, Sylvia R. Munter, is born in Brooklyn.

January 31, 1913 David Madoff becomes a U.S. citizen.

1930 Madoff family resettles in the Bronx. David works as a tailor at a clothing store, and Ralph works as an assistant manager at a jewelry store.

1932 Bernie Madoff's parents, Ralph and Sylvia, marry at the height of the Great Depression. Ralph is 24 and Sylvia is 20. On the marriage license, Ralph's occupation is listed as "Credit" and Sylvia's as "None."

June 14, 1934 Bernie Madoff's older sister, Sondra, is born.

April 29, 1938 Bernard Lawrence Madoff is born.

May 18, 1941 Bernie's future wife, Ruth Alpern, is born.

October 19, 1945 Bernie Madoff's brother, Peter, is born.

April 1946 The Madoff family moves from Belmont Avenue in Brooklyn to Laurelton, Queens.

1952 Bernie Madoff enters Far Rockaway High School.

April 1953 J. Ezra Merkin is born.

Summer 1954 Bernie Madoff works as a lifeguard at Jacob Riis Park, making $8.78 a day.

Summer 1955 Bernie Madoff works as a lifeguard at Rockaway Beach.

June 26, 1956 Bernie Madoff graduates from Far Rockaway High School.

Summer 1956 Bernie Madoff works as a lifeguard at Jacob Riis Park, making $10 a day.

Fall 1956 Bernie Madoff attends the University of Alabama for the first semester of his freshman year.

October 1956 Harry M. Markopolos is born in Erie, Pennsylvania.

1957 Bernie Madoff transfers to Hofstra College.

1958 The Fifth Avenue Synagogue is founded.

November 29, 1959 Bernie Madoff marries Ruth Alpern.

1960 Carl Shapiro meets Bernie Madoff.

March 29, 1960 Bernie Madoff qualifies as a general securities representative and a general securities principal.

June 6, 1960 Bernie Madoff graduates from Hofstra College with a degree in political science. He then attends Brooklyn Law School for a year, but drops out to begin his business.

September 12, 1960 Bernie Madoff is commissioned as a U.S. Army Reserve second lieutenant in the Infantry branch.

November 1960 Bernard L. Madoff Investment Securities is founded.

1962 Accountants Frank Avellino and Michael Bienes begin to raise money for Madoff. They promise investors returns of 13.5 to 20 percent.

1963 Neil and Constance Friedman are introduced to Bernie Madoff by their friend, Jerry Horowitz.

August 6, 1963 Gibraltar Securities, a broker-dealer firm registered in Sylvia Madoff's name, is investigated by the SEC.

November 7, 1963 Bernie Madoff is discharged from the Army Reserve.

January 23, 1964 Gibraltar Securities agrees with the SEC to shut down and not face legal action.

March 21, 1964 Bernie and Ruth Madoff's first son, Mark David Madoff, is born.

April 28, 1966 Bernie and Ruth Madoff's second son, Andrew Howard Madoff, is born.

1967 Peter Madoff graduates Fordham Law School, joins his brother's firm, and has his first child, Shana.

1968 Future Madoff middleman Robert Jaffe marries Ellen Shapiro, daughter of Carl Shapiro.

1971 The NASDAQ market is introduced. Carl Shapiro makes a fortune selling Kay Windsor to Vanity Fair.

July 1972 Bernie Madoff's father, Ralph, dies.

1974 Saul Alpern retires. His accounting firm is taken over by Avellino and Bienes.

December 1974 Bernie Madoff's mother, Sylvia, dies.

1977 Bernie and Ruth Madoff buy a 55½-foot Rybovich boat for $462,000. The boat is christened *Bull*.

1979 Bernie and Ruth Madoff buy a $3 million beach house in Montauk.

May 1981 Harry Markopolos graduates from Loyola College in Baltimore with a B.S. in accounting.

1984 Bernie and Ruth Madoff buy their $7 million Manhattan penthouse.

1983 Walter Noel founds his investment company. Madoff Securities International in London opens.

January 1984 Bernie Madoff becomes a member of the board of governors of the National Association of Securities Dealers.

1985 Attorney J. Ezra Merkin founds Gabriel Capital LP.

February 1985 Brokerage firm Cohmad Securities is founded by Bernie Madoff and his longtime friend Maurice J. Cohn.

1986 Mark Madoff joins the family firm after graduating from the University of Michigan.

1989 Walter Noel and Jeffrey Tucker make their first investment with Madoff.

Andrew Madoff joins the family firm after graduating from the University of Pennsylvania.

May 20, 1989 Andrés Piedrahíta marries Walter Noel's daughter, Corina.

1990 Bernie Madoff serves his first year as nonexecutive chairman of NASDAQ. He will serve again in 1991 and 1993.

Ezra Merkin begins investing with Madoff.

November 17, 1992 Avellino and Bienes are accused of selling $441 million in illegal securities to thousands of investors over three decades.

1993 Peter Madoff serves as vice chairman of the NASD.

November 22, 1993 Avellino and Bienes are forced to shut down and pay a fine of $350,000.

1994 Ezra Merkin becomes the chairman of Yeshiva University's investment committee.

March 1994 Bernie and Ruth Madoff buy their home in Palm Beach.

1996 Bernie and Ruth Madoff join the Palm Beach Country Club, paying the $300,000 entry fee.

Bernie Madoff begins serving on Yeshiva University's board of trustees.

Ruth Madoff executive-edits a cookbook, *Great Chefs of America Cook Kosher*.

1997 Andrés Piedrahíta becomes a partner at Fairfield Greenwich; soon after, he and his wife, Corina, move to London.

1998 Mark and Andy Madoff become directors of Madoff Securities International in London and take stakes in the business.

Bernie and Ruth Madoff establish the Madoff Family Foundation.

November 1999 Frank Casey asks Harry Markopolos to reverse-engineer the Madoff strategy. Within minutes, Markopolos decides it's a fraud.

2000 Bernie Madoff becomes chairman of Yeshiva University's Sy Syms School of Business.

May 2000 Harry Markopolos makes the first submission of his Madoff investigation to the SEC in Boston.

June 2000 Credit Suisse urges its customers to withdraw investments from Madoff's fund.

May 2001 Michael Ocrant of *MARHedge* publishes his story on Madoff, followed by Erin Arvedlund of *Barron's*.

October 2001 Harry Markopolos submits his Madoff investigation to the SEC a second time.

June 19–29, 2002 Harry Markopolos meets with European money managers enamored with Madoff's investment fund.

October 2002 *Vanity Fair* publishes "Golden in Greenwich," an article on the beautiful Noel daughters.

November 25, 2002 Peter Madoff's son, Roger, is diagnosed with leukemia.

December 17, 2002 Harry Markopolos, wearing white gloves, attempts to deliver his Madoff report to New York State Attorney General Eliot Spitzer.

March 2003 Société Générale puts Madoff's fund on its internal blacklist.

November 4, 2005 Harry Markopolos emails Meaghan Cheung at the SEC's New York office a memo titled "The World's Largest Hedge Fund Is a Fraud."

December 2005 Bernie Madoff coaches executives at Fairfield Greenwich on how to answer questions from SEC investigators.

January 4, 2006 The SEC formally opens an investigation into Madoff based on Harry Markopolos's report.

April 15, 2006 Roger Madoff succumbs to leukemia at the age of 32.

May 19, 2006 Bernie Madoff lies under oath about how he runs his investment advisory.

November 21, 2007 The SEC officially closes its Madoff case, finding "no evidence of fraud."

February 15, 2008 Carl Shapiro celebrates his ninety-fifth birthday at Club Collette in Palm Beach.

April 2008 Fairfield Greenwich executives discuss the risk that Madoff will "blow up."

May 2008 Fairfield client Unigestion asks point-blank if Madoff was making the trades and holding the assets he claimed.

August 2008 JPMorgan Chase grows suspicious of Madoff, withdraws its investments.

November 25, 2008 Ruth Madoff transfers $5.5 million from her account with Cohmad Securities into her personal bank account.

December 8, 2008 Bernie Madoff lashes out at Fairfield Greenwich for failing to replenish its dwindling investments.

December 9, 2008 Bernie Madoff confesses his fraud to his brother.

December 10, 2008 Ruth Madoff withdraws $10 million from Cohmad Securities.

Bernie Madoff confesses to his sons, saying his fund was "basically a giant Ponzi scheme."

December 11, 2008 The FBI arrests Bernie Madoff and charges him with criminal securities fraud.

March 12, 2009 Bernie Madoff pleads guilty to eleven criminal charges and is sent to jail pending sentencing.

June 29, 2009 Bernie Madoff is sentenced to 150 years in prison.

Notes

Introduction

Carl Shapiro's ninety-fifth birthday party was a night to remember. I interviewed four attendees at the party on a not-for-attribution basis. The *Palm Beach Daily News* captured the celebration in the article "He's 95 and He'll Sing If He Wants To," published February 20, 2008.

The private performance of the Israeli Philharmonic string quartet in Howard Kessler's living room was reported by newyorksocialdiary .com in December 2007, which included photographs and a video recording of the quartet in Kessler's home.

Abe Gosman's financial devastation was outlined in "Ex-socialite Lin Gosman Indicted, Charged with Hiding Assets," *Palm Beach Post*, November 7, 2008.

I was present in the courtroom when Bernie Madoff pled guilty to all eleven counts charged by the government and heard him read his prepared statement to his victims.

1. The Struggler

My opening description of Laurelton comes from the stories and memories of countless people who knew and lived near young Bernie Madoff. Many of them, such as Elsa Levine, are introduced in the chapter while other friends and classmates wished to remain anonymous. Carol Solomon Marston and her website, www.farrockaway.com, were tremendously helpful with my depictions of Far Rockaway and Laurelton as they existed over fifty years ago. I also spent time in Laurelton.

The story of Stephen Richards and Bob Roman meeting Bernie Madoff and Ruth and Joan Alpern comes from an interview I conducted with Richards. Bob and Joan Roman declined to comment.

The description of Madoff as a boy is from interviews I conducted with several of his and Peter's friends, family members (who did not wish to be quoted) and classmates, as well as several photographs I found during the course of my research. These photos include an enlarged fifth-grade class picture from P.S. 156, Madoff's senior picture from Far Rockaway High School's class of 1956 yearbook, and a photo from his freshman year at the University of Alabama in the school's 1957 yearbook. Both the P.S. 156 class photo and the Far Rockaway High School yearbook were given to me by Madoff's childhood friend, Jay Portnoy. Madoff's photo from the University of Alabama comes from "An Unlikely Antihero: Bernie Madoff at the Capstone," a posting on the University of Alabama's library blog.

My interview with Madoff's childhood neighbor Donny Rosenzweig provided me with the details about Bernie's sprinkler business, as well as a description of how much time Bernie and Ruth spent together. Cynthia Arenson also shared her memories of Ruth with me.

Bernie's dubious claim that he fought his way out of an impoverished childhood can be found in, among other places, "The Monster Mensch," *New York*, February 22, 2009.

My information on Bernie Madoff's paternal grandparents comes from several immigration and census documents. The most useful were David Solomon Madoff's petition for naturalization, filed on January 31, 1913, with the U.S. District Court for the Middle District of Pennsylvania (Scranton). I also found the U.S. Census documents for the Madoff family in 1920 when they lived in Scranton, Pennsylvania, and in 1930 when they lived in the Bronx. Census documents list the names of each family member, the languages they knew, their occupations, country of origin, and the value of their home or monthly rent. This information was invaluable when it came to painting a picture of Bernie Madoff's upbringing and family history.

Material gathered about Bernie Madoff's maternal grandparents, the Muntners, was also obtained through the U.S. Census, specifically the Muntners' 1930 Lower East Side census listing. The census confirms Harry Muntner's occupation as the proprietor of the bathhouse at 9 Essex Street. Martin Schames, owner of M. Schames & Son Paint on the block where the Muntners' brownstone once stood, was able to recall the days of his childhood when the bathhouse was in operation. Joyce Mendelsohn, author of *The Lower East Side Remembered and Revisited*, was also a tremendous help in this area of research. She put me in touch with Martin Schames and was always available to offer her expertise.

A copy of Sylvia Muntner and Ralph Madoff's marriage certificate was obtained from the archives of the New York City clerk. Ralph and Sylvia Madoff moved from Belmont Avenue in Brooklyn and bought their Laurelton house in April 1946, according to property records documenting the sale. Confirmation of the Madoffs' address in Laurelton comes from Queens County phone books between 1947 and 1956.

Helen Freedman told me about Bernie's personality and the games he would play as a child.

The theme of Bernie as a lackluster student recurred in multiple interviews, especially with Elsa Levine, Marcia Mendelsohn, and others who asked to remain anonymous. These individuals also provided details on the Ravens, Elliot Olin, and the culture of P.S. 156.

During our interviews, Elsa Levine recounted Bernie's crush on her and Marcia Mendelsohn detailed Bernie's inferiority complex and her young romance with him.

Jay Portnoy's memories of commuting with Madoff and his friends to Far Rockaway High School are in "Growing Up with the Ponzi

King," *Saratogian News*, February 13, 2009, and were recounted to me by Portnoy himself on March 9, 2009. The anecdote about Bernie's fake book report also comes from Portnoy. Background on Bernie's high school years was also found in "Bernie Madoff Had a Job Saving Lives, Now He's Killing Dreams," *Daily News*, December 21, 2008. This article also had important information about Bernie's years on the Far Rockaway swim team. I interviewed Fletcher Eberle, the captain of Bernie's swim team, about Bernie during competitions and also about Bernie's job as a lifeguard.

Joyce Lahn told me about cutting school at Far Rockaway to hang out at the beach or Hal's Luncheonette.

Irene Shapiro wrote to me about Laurelton's culture of material wealth in a series of emails between February 1 and February 3, 2009. She and several other residents of Laurelton provided a plethora of details about the differences between Laurelton and the Five Towns.

I used Far Rockaway High School's commencement program to confirm the date of Madoff's graduation. The program was scanned and posted online by Carol Solomon Marston at www.farrockaway .com. Jay Portnoy provided me with the yearbook from Madoff's graduating class, which listed the college or other destination for each student.

My interview with Martin Schrager, Bernie's roommate at the University of Alabama, and the history of the University of Alabama's Hillel House, found at www.bamahillel.org, contributed to my depiction of Bernie's semester in the South. "Bernie Madoff, Frat Brother," *New York Times*, January 16, 2009, was the springboard for my research on Bernie's brief stint in Alabama.

2. The Plan

Jay Portnoy described his frightening drive with Bernie Madoff in our interview. He also penned his memories of the episode in the February

13, 2009, article "Growing Up with the Ponzi King," written for the *Saratogian News*.

For a March 22, 2009, article in the *Times* of London titled "Life inside the Weird World of Bernard Madoff," Bill Nasi recalled a conversation when Ralph Madoff told him never to invest in the stock market and never to let greed get into his psyche.

My team of researchers and I scoured SEC records and found no evidence of Ralph or Sylvia Madoff ever holding a stockbroker's license. We were able to find records of Sylvia Madoff with two brokerages registered in her name, Gibraltar Securities and Second Gibraltar Corporation.

My interview with Stephen Richards revealed that Bernie Madoff worked as a stockbroker before he was licensed. Richards said that he cleared trades through his father, who was never licensed as a stockbroker.

The SEC ordered Sylvia Madoff to shut down her brokerage business after she failed to file at least one year's worth of financial statements. I found a detailed account of the SEC beginning its investigation of Gibraltar Securities in an August 6, 1963, edition of *SEC News Digest*. The outcome of the SEC's action, when Gibraltar agreed to shut down and not face prosecution, is reported in a January 23, 1964, edition of *SEC News Digest*.

The 1956 tax lien on the Madoff home was retrieved from the real estate records kept by the Queens County clerk.

Information on Madoff's graduating class from Hofstra College came from a notice in the *New York Times* on June 6, 1960.

Details about Bernie and Ruth's wedding were reported in "Madoff's World," *Vanity Fair*, April 2009. They were expanded on and confirmed to me by Madoff's close friends, who requested anonymity.

"The Talented Mr. Madoff," *New York Times*, January 24, 2009, provided details about Bernie and Ruth's first apartment together in

Bayside, Queens. I visited the apartment building myself to be able to more fully describe their first home.

Madoff's military records from the years he served as a U.S. Army Reserve second lieutenant were obtained courtesy of U.S. Army Public Affairs.

Confirmation of when Madoff passed his General Securities Representative exam and General Securities Principal exam came from his broker report, issued by the Financial Industry Regulatory Authority (FINRA).

The rise of Bernie Madoff's market-making empire and the creation of NASDAQ injected more competition into an arena once dominated solely by the New York Stock Exchange. For background on the growth of Madoff's business, I consulted "Madoff Created an Air of Mystery," *Wall Street Journal*, December 20, 2008, and "The Monster Mensch," *New York*, February 22, 2009. Another valuable resource was the book *What Goes Up* by Eric J. Weiner. In it, Madoff offers very simple commentary on how NASDAQ was created.

Carl Shapiro took on Bernie Madoff as his protégé nearly fifty years ago, at the height of his clothing company's incredible success. I was able to look into Shapiro's life during his midforties by finding a January 27, 1957, *New York Times* article, " 'Work Horse' Dress Builds a $22,000,000 Business: Big Concern Built on Cotton Gowns Started with Father Runs Business Alone." I reconstructed Shapiro's upbringing and family history by finding his family's listing in the 1930 U.S. Census.

Shapiro's story about how he met Madoff was detailed in not-for-attribution interviews with friends and "Palm Beacher Calls Bernie Madoff Arrest 'Knife in the Heart,' " *Palm Beach Daily News*, December 16, 2008. Several Boston residents gave me useful background on Carl Shapiro's reputation and charitable contributions.

The Sunny Oaks Hotel and Cottages in Woodridge, New York, was the first place Bernie Madoff found clients to fund his investment business. *Bloomberg* ran a story on January 29, 2009, titled "Madoff's

Tactics Date to 1960s, When Father-in-law Was Recruiter," which jump-started my investigation into the Sunny Oaks and led me to conduct several interviews with Cynthia Arenson and conduct additional research.

Frank Avellino and Michael Bienes had a long and complicated relationship with Bernie Madoff. PBS's *Frontline* published the transcripts of its extraordinary May 13, 2009, interview with Bienes that helped me to explain the history shared among the three men. "Former Madoff Associate Michael Bienes Breaks His Silence," *Sun Sentinel* (Fort Lauderdale), March 9, 2009, was a useful resource, as was "How Bernie Did It," *Fortune*, April 30, 2009.

Bernie's astronomical stock offering by his fledgling market-making business was taken from an *SEC News Digest* column on March 30, 1962. This article also listed the address of Bernie's firm, helping me to piece together where he worked decades before he moved into the Lipstick Building.

During our interview on January 29, 2009, Stephen Richards provided me with the story of how he invested $100,000 with Madoff.

3. The In Crowd

Palm Beach is one of the most beautiful communities in the world, yet its glamorous appearance belies an ugly history of social segregation. I learned of the city's heritage from many residents I met with during my week's stay in Palm Beach in February 2009.

David Neff, the owner of the Palm Beach clothing store Trillion, told me about Bernie's outlandish shopping habits during our interview on February 17, 2009.

Descriptions of Bernie and Ruth's inseparable relationship, Ruth's outgoing personality, and Bernie's behavior in Palm Beach were contributed by two intimate friends with the Madoffs of several decades. Both wish to remain anonymous.

Details on Madoff's boat were initially found in "Madoff Enjoyed $50 Pedicures, 9.8 Handicap, Boat," *Bloomberg*, December 17, 2008, and subsequently confirmed by my researchers.

Facts surrounding the Madoffs' house in Palm Beach, including its value and date of purchase, were found in the Palm Beach County Property Appraiser Public Access System, a service that can be accessed online at www.pbcgov.com/papa.

Details of Michael Bienes's lavish lifestyle in Florida and England stem from "Madoff's Man," *Broward-Palm Beach New Times*, January 20, 2009. I looked at several photographs from the archives of the *Miami Herald* featuring Michael and Dianne Bienes displaying their home's art collection. The photos were taken on November 15, 1999, in the Bieneses' Fort Lauderdale mansion.

Facts surrounding the SEC's action against Avellino and Bienes in 1992 came from documents filed with the SEC complaint *SEC v. Avellino & Bienes*, Frank J. Avellino and Michael S. Bienes, 92 Civ. 8314, JES, SDNY. More facts and analysis surrounding the case came from "Look at Wall St. Wizard Finds Magic Had Skeptics," *New York Times*, December 12, 2008, and " '92 Ponzi Case Missed Signals about Madoff," *New York Times*, January 16, 2009.

To capture the mystery surrounding the 1992 SEC investigation into Avellino and Bienes, I quoted from Randal Smith's December 16, 1992, article in the *Wall Street Journal*, "Wall Street Mystery Features a Big Board Rival."

Michael Bienes described his explosive confrontation with Bernie when the SEC concluded its work on PBS's *Frontline*, May 13, 2009.

During our interview on January 29, 2009, Stephen Richards stated that he kept his money invested with Madoff even after the Avellino and Bienes debacle.

For an overview of the relationship between Bernie Madoff and Robert Jaffe, I consulted, among other sources, "Scandal Sullies Robert Jaffe as Feds Probe Ties to Bernard Madoff," *Palm Beach Post*, December 20, 2008; "Madoff's World," *Vanity Fair*, April 2009; and "Access to Bernard Madoff Made Middleman Robert Jaffe a 'Superstar,'" *Boston Globe*, December 21, 2008.

I took in the atmosphere of the Palm Beach Country Club during my visit in February 2009. Society writer Lawrence Leamer's piece for *Boston Magazine*, "Dispatch: Reversal of Fortune," provided me with details about the Club's wealthy membership.

I owe Richard Rampell many thanks for discussing his impressions of Palm Beach society with me. My interview with him on March 16, 2009, covered Robert Jaffe, Carl Shapiro, and Rampell's puzzling encounter with Bernie Madoff.

4. Friends and Enemies

Harry Markopolos pursued Bernard Madoff for nearly a decade. Markopolos detailed his military background in a Q&A published in "Inside the Madoff Scandal: Army Special Operations Training for Whistleblower," *Boston Progressive Examiner*, March 20, 2009.

Frank Casey recounted his trip to see René-Thierry Magon de la Villehuchet during an interview with the Fox Business Network on December 19, 2008.

Background on de la Villehuchet comes from interviews and research compiled by my assistant, Nathalie Mattheiem. She interviewed British photographer Guy Gurney, de la Villehuchet's occasional sailing partner. Mattheiem also interviewed Marie-Monique Sterkel, a member of his tight circle of friends, who spent time with him at his chateau. Jacques Puisségur-Ripet, a sailing comrade, and Jean-François Hénin, once known in France as the "Mozart of finance," were other de la Villehuchet's friends interviewed by Mattheiem.

More on de la Villehuchet's background came from "Le Suicide pour l'Honneur de Thierry de la Villehuchet," *Paris Match*, December 30, 2008. Also of value were "Thierry S'est Senti Trahi par Bernard Madoff," *Rue89.com*, December 27, 2008, and "Madoff Fund Operator De La Villehuchet Found Dead," *Bloomberg*, December 23, 2008.

Markopolos declined my interview requests. I interviewed Michael Ocrant of *MARHedge*, whom Markopolos cites as a member of his team investigating Madoff.

The narrative describing how Markopolos began his investigation into Madoff's fraud comes from the events outlined in Markopolos's written and delivered testimony to the U.S. House Committee on Financial Services, February 4, 2009.

Mathematician Dan diBartolomeo was the first person to independently verify Harry Markopolos's findings about the Madoff fraud. I interviewed diBartolomeo to get his firsthand account of what happened when he reviewed Markopolos's data.

Harry Markopolos's degree and dates of attendance at Loyola College in Baltimore were confirmed by Courtney Jolley, director of Loyola Public Affairs. His master's degree in finance from Boston College was confirmed by Ed Heyward, BC Department of Public Relations.

Markopolos's childhood and high school friends recalled his years growing up and attending Cathedral Preparatory School in "Erie Man Blew Whistle on Madoff," *Erie Times-News*, December 20, 2008. Markopolos outlined his professional experience during his testimony to the U.S. House Committee on Financial Services, February 4, 2009.

I interviewed John Pohlad, managing director and CEO of Marquette Asset Management, about the competition he received from Bernie Madoff. Pohlad was my primary source for Madoff's impact on the Twin Cities. Another valuable resource came from "Madoff Does Minneapolis," *Fortune*, January 22, 2009. *Fortune*'s piece helped me describe the impact Mike Engler had in Minnesota.

My description of the Fifth Avenue Synagogue, its leaders, and its congregation comes from three not-for-attribution interviews with synagogue members.

Biographical information on Ezra Merkin came from *New York* magazine's superb article "The Monster Mensch," published February 22, 2009, as well as "Merkin Intimidated Co-Op Board While Building Funds Madoff Lost," *Bloomberg*, January 9, 2009. Merkin background and a list of the celebrated American figures who have called Merkin's building home came from Michael Gross's book, *740 Park: The Story of the World's Richest Apartment Building*. William Ackman and Michael Steinhardt also contributed color and stories about their friendship with Merkin at the YIVO panel discussion I attended, "Madoff: A Jewish Reckoning," held on January 15, 2009.

Merkin was erudite from a very young age. To exemplify this point, I quoted a passage from Daphne Merkin's book, *Enchantment*, a work of autobiographical fiction published in 1987.

Merkin's first money manager before Bernie Madoff, Victor Teicher, has a past that includes jail time for insider trading, reported in "In Bernie Madoff Investigation, Financial Adviser Dodges Questions," *Daily News*, April 6, 2009.

Merkin and Madoff's involvement with Yeshiva University is chronicled in "After Madoff: Charities Still Picking Up the Pieces," *NonProfit Times*, March 4, 2009.

New York Attorney General Andrew Cuomo's complaint against J. Ezra Merkin and Gabriel Capital Corporation, *New York v. Merkin*, 09-450879, New York State Supreme Court (Manhattan), provided details on the fees Merkin would charge clients simply to funnel their money to Bernie Madoff. New York University's lawsuit against Merkin, *New York University v. Ariel Fund Ltd.*, 8603803/2008, New York State Supreme Court (Manhattan), helped me to understand allegations that Merkin concealed Bernie Madoff's role in his clients' investments.

Merkin has denied all the charges. Merkin's attorney, Andrew Levander, did not respond to my interview request, nor did his sister, Daphne.

Nobel laureate Elie Wiesel was one of the most notable Madoff victims. He described how he was lured into the scheme during a panel discussion arranged by *Portfolio* on February 26, 2009.

Ruth Madoff was the executive editor of the cookbook *Great Chefs of America Cook Kosher*. She claimed to have put together much of the book herself, but Karen MacNeil told the *New York Times* in the January 14, 2009, article "A Madoff Cookbook Has a Secret, Too" that she was paid to write the cookbook in its entirety. I also interviewed a close friend of Ruth's about the subject on a not-for-attribution basis.

Harry Markopolos's submission of his Madoff investigation to Grant Ward, director of enforcement for the SEC in New England, was recounted in his written testimony to the U.S. House Committee on Financial Services, February 4, 2009. He also details the encounter in his prepared presentation and in notes and emails he submitted as part of that testimony. Grant Ward told me he had no comment on the matter.

5. The Friendly Company

Mark Madoff's quote is in "Family Influence: The Madoff Dynasty," *FinanceTech*, July 7, 2000. The dates of when each Madoff family member joined Bernard L. Madoff Investment Securities were confirmed by their broker profiles issued by FINRA.

The atmosphere of Bernie Madoff's workplace was described to me in interviews with six market-making traders, four company drivers, three information technology employees, one secretary, and one intimate friend.

Eleanor Squillari's "Hello, Madoff!," written for the May 2009 issue of *Vanity Fair*, contributed several stories about the bizarre and befuddling office that was Bernard L. Madoff Investment Securities. Her anecdotes were confirmed and often elaborated on by Elaine Solomon during the course of our conversations.

Several articles provided insight into payment for order flow and how Madoff's business operated. These included "Madoff Created an Air of Mystery," *Wall Street Journal*, December 20, 2008, and "How Bernie Did It," *Fortune*, April 30, 2009.

Bill Nasi chronicled the strange incidents and eccentricities he saw while working as a messenger for Bernie Madoff in "Life inside the Weird World of Bernie Madoff," *Times* (London), March 22, 2009. In the article, he also described the relationship between Bernie and Peter Madoff, recalling an intense screaming match between the two.

During an interview I conducted with her, Jan Chasen, one of Peter Madoff's childhood friends from Laurelton, told me that Peter was more erudite than his brother, Bernie.

Andy's high school senior yearbook quote was found in the 1984 Roslyn High School yearbook.

In its January 18, 2009, article "Private Clubs: Hideouts of the Rich and Shameless," the *New York Post* revealed that Mark Madoff was once a member of the Core Club. Traders from the market-making operation on the nineteenth floor gave me details about the sons and their personalities over the course of numerous interviews.

Mark casually shopping the J. Crew website in the midst of a market crash and Andy's disastrous foray into trading on his own were described to me by traders whose interviews were not for attribution.

Press sources for color about Shana Madoff come from *New York* magazine's fashion profile of her on August 14, 2004, "Extreme Brand Loyalty," and "Shana Madoff's Ties to Uncle Probed," *Wall Street Journal*, December 22, 2008.

Near the beginning of my research process, Donny Rosenzweig told me about Madoff's apparent generosity when he gave Rosenzweig entrée into his fund.

6. The Seventeenth Floor

Company driver Clive Brown revealed in an interview with me that it was primarily his responsibility to hand-deliver daily checks for Norman Levy, in amounts often totaling millions of dollars.

Repeated conversations with spokesmen for JPMorgan Chase never resulted in an explanation of why Levy, who was in the real estate business, was given an office at the bank's Park Avenue headquarters, or if JPMorgan Chase's compliance systems detected the deposits.

I interviewed Madoff employees and Madoff friends about Norman Levy. Several details on Levy's life came from "Madoff's World," an excellent piece of reporting published in the April 2009 edition of *Vanity Fair.* More on Levy's relationship with Madoff was provided by Levy's son, Francis Levy, during an interview on the Fox Business Network on December 15, 2008. Norman Levy passed away in 2005.

Descriptions of the infamous seventeenth floor, Annette Bongiorno, Frank DiPascali, and the employees who worked under both managers came from interviews with Bob McMahon, Ken Hutchinson, and not-for-attribution conversations. McMahon and Hutchinson worked for Madoff's IT department and often expressed extreme curiosity regarding the secrecy surrounding seventeen.

Several points about Annette Bongiorno's life and role on the seventeenth floor came from "Madoff Key Aide Bongiorno Recruited Her Neighbors as Investors," *Bloomberg*, February 14, 2009.

Details of Frank DiPascali's personality came from a host of interviews with Madoff employees, including Ken Hutchinson. Valuable information about DiPascali also came from "Madoff Investigation Involves Bridgewater Man Who Worked for Financier," *Bloomberg*, January 23, 2009.

IT employees George Perez and Jerry O'Hara never responded to my phone calls about their work on the investment advisory's AS400 server.

Bernie Madoff could have sold his market-making business as late as 2002 for $1 billion, according to "The Monster Mensch," *New York*, February 22, 2009.

7. The Black Box

Cynthia Arenson and Stephen Richards described for me how their investments with Madoff changed their lives; Arenson bulldozed the Sunny Oaks to build a fabulous vacation home, and Richards sold his furniture business and retired to Boca Raton.

Madoff's contagion spread to the West Coast, embodied by the stories of business managers Gerald Breslauer and Stanley Chais. Both men promoted Madoff's fund to their wealthy clients, from the gentry of Beverly Hills to famous Hollywood names.

Information concerning Stanley Chais and his history of investing with Madoff comes from the complaint filed by Madoff trustee Irving H. Picard against Chais, his companies, and family members on May 1, 2009. Another look into Chais's astounding Madoff profits and lack of suspicions comes from a complaint filed by the SEC against Chais on June 22, 2009.

Madoff's presence in Hollywood was reported in "Madoff's Hollywood Connection," *Portfolio*, March 2009. "Hollywood Figures Snared in Bernard Madoff's Alleged Fraud," *Los Angeles Times*, December 17, 2008, also outlined Madoff's impact on the West Coast.

An extensive interview I had with Michael Ocrant was invaluable for my description of how he began his 2001 investigation into Madoff's investment advisory. I also used televised interviews of Casey describing his meeting with Ocrant. Erin Arvedlund recounted how she started her investigation for *Barron's* in an essay published by *Portfolio* on December 17, 2008. She did not return calls for comment.

The story of Mark Madoff standing on a desk in the middle of the trading floor and assuring employees that the *MARHedge* and *Barron's* stories were untrue was recounted to me by several traders. I also referred to "Son's Roles in Spotlight," *Wall Street Journal*, January 24, 2009.

Harry Markopolos told Boston's WBUR Radio about the fruit fly prank he pulled growing up in Erie, Pennsylvania, during his interview with the station on April 21, 2009. Markopolos covered a range of topics during the WBUR interview, including how he began fearing for his life after his second submission to the SEC in October 2001. The events surrounding the 2001 submission are explained in Markopolos's written testimony to the U.S. House Committee on Financial Services, February 4, 2009, and in his SEC submission itself.

8. The World at His Feet

Sherry Shameer Cohen's impressions of Fairfield Greenwich and the Noel family served as the basis of the opening narrative in chapter 6. Cohen first caught my attention when she briefly appeared on PBS's *Frontline* on May 13, 2009. She was given only a few minutes of airtime, but her extended interview transcripts released by PBS and the postings from her blog, "Metro Journalist," convinced me to interview her more extensively. During our conversation, Cohen helped me understand the remarkable transformation of Walter Noel's tiny hedge fund after it invested with Bernie Madoff. She also shared her feelings about the Noel daughters and sons-in-law as she knew them during her eleven-year tenure at Fairfield Greenwich.

Several articles provided integral facts and stories about the history of the Noel family. Among the pieces I referred to most often were "Madoff Scheme Kept Rippling Outward, across Borders," *New York Times*, December 19, 2008; "Golden in Greenwich," *Vanity Fair*, October 2002; and "Greenwich Mean Time," *Vanity Fair*, April 2009.

I spent several weeks negotiating with Walter Noel's spokesperson and lawyers to conduct an interview for this book. Unfortunately, his company lawyers prevented him from doing so, but they did respond to several allegations included in this chapter.

Andrés Piedrahíta was interviewed by the *Wall Street Journal* for a March 31, 2009, article titled "The Charming Mr. Piedrahita Finds Himself Caught in the Madoff Storm." The *Journal* documented the lavish lifestyle he shared with his wife, Corina Noel. The international edition of Colombia's *Semana* magazine provided more biographical information on Piedrahíta in the article, "The Colombian in the Madoff Scandal," December 25, 2008.

David Giampaolo described the questionable investment presentation given to him and a friend by Piedrahíta in "Fairfield Extended Madoff's Reach," *Wall Street Journal*, December 19, 2008. Messages left in London for Giampaolo to conduct a follow-up interview were not returned.

Jeffrey Tucker's attempt to conduct due diligence with Madoff in the spring of 2001 was documented in internal Fairfield Greenwich emails found in the administrative complaint filed by the Massachusetts Secretary of the Commonwealth against Fairfield Greenwich Advisors LLC and Fairfield Greenwich (Bermuda) LTD on April 1, 2009.

Details on the Madoffs' residence in Cap d'Antibes came from "Madoff's Three-bedroom Riviera Retreat Belied Ponzi Scheme Role," *Bloomberg*, January 9, 2009. Elaine Solomon was also able to tell me some tidbits about the Madoffs' stays at the Hotel Plaza Athénée in Paris.

Banco Santander's losses were outlined in "Giant Bank in Probe over Ties to Madoff," *Wall Street Journal*, January 13, 2009.

Biographical information about Sonja Kohn came from "Profile: Austrian Bank Executive Sonja Kohn," *Financial Times*, December 22, 2008, and "Sonja Kohn Wooed Bernard Madoff Billions with Medici 'Fantasy,' " *Bloomberg*, February 19, 2009.

Facts on Madoff's London office, including its role in the Ponzi scheme, were compiled in "Madoff Used U.K. Office in Cash Ploy, Filing Says," *Wall Street Journal*, March 12, 2009. Madoff's money-laundering charges involving transactions wired to his London office were found in the criminal information filed against Bernie Madoff by the Southern District of New York on March 10, 2009. On March 13, 2006, Madoff pled guilty to these charges.

Bernie Madoff's London assistant, Julia Fenwick, offered detailed stories on her boss's extravagances and eccentricities in "I Just Can't Live With That Camera—It's Not Square . . . Inside the Bizarre World of £30bn Pyramid Schemester Bernie Madoff," *Daily Mail* (London), January 3, 2009.

Harry Markopolos explained the events of his 2002 trip to Europe in his written testimony to the U.S. House Committee on Financial Services, February 4, 2009. He provided several documents not discussed during his testimony, which outlined who he traveled with, where he stayed, and what meetings he attended. My researchers and I perused these documents to reconstruct his trip as accurately as possible. Markopolos also covered the events of his European trip in his April 21, 2009, interview on WBUR Radio.

Credit Suisse's episode where their due diligence led to a withdrawal of all Madoff investments is featured in "Credit Suisse Urged Clients to Dump Madoff Funds," *Bloomberg*, January 7, 2009. Société Générale's adding Bernie Madoff to its blacklist in March 2003 is detailed in "European Banks Tally Losses Linked to Fraud," *New York Times*, December 16, 2009.

9. See No Evil

Harry Markopolos described his attempt to deliver an envelope to New York Attorney General Eliot Spitzer during his testimony to the U.S. House Committee on Financial Services, February 4, 2009. The envelope contained all the documents of his investigation into Bernie Madoff. I interviewed Governor Spitzer about the event he and Markopolos attended, and he told me he never received the envelope. Spitzer (as state attorney general) was attending a panel discussion at the John F. Kennedy Presidential Library in 2002. I viewed a video of the event.

Genevievette Walker-Lightfoot's suspicions in 2004 that Madoff's investment advisory was a fraud were reported in "Staffer at SEC Had Warned of Madoff," *Washington Post*, July 2, 2009.

Elaine Solomon vividly remembered several episodes when the SEC came to the Lipstick Building to investigate Madoff. She described the SEC investigators looking like they were still children and handing out résumés.

My interview with Neil and Constance Friedman provided the details for Jerry Horowitz's fiftieth anniversary celebration, where everyone toasted Bernie for their good fortune. Jerry's son, Mitchell Horowitz, did not return requests for comment. The Friedmans also recounted the sky-high profits from Madoff at the time of the party and their move to a beautiful home in Palm City, Florida.

Michael Bienes's memories of Jerry Horowitz were taken from the transcript of his extended interview with PBS's *Frontline* on May 13, 2009. I used "Madoff's Accountant 'Was Taken' Like Other Investors, Man Says," *Bloomberg*, December 20, 2008, to assist my description of Horowitz's early business with Madoff.

David Friehling and his background were covered in "Arrest Stuns Friehling's Friends, Colleagues," *Journal News*, March 22, 2009. The criminal complaint filed on March 18, 2009, by the Southern District of New York against Friehling provided information on the history of Friehling's business with Madoff, as well as the Generally Accepted Accounting Standards with which he allegedly failed to comply. The allegations that Friehling misrepresented to the American Institute of Certified Public Accountants that his firm no longer performed audits also comes from the text of the complaint.

I read through multiple internal emails from Fairfield Greenwich dated September 2005, when Chief Financial Officer Dan Lipton produced a response for a client who asked if Fairfield properly vetted Friehling and Horowitz. The emails were found in the administrative complaint filed by the Massachusetts Secretary of the Commonwealth against Fairfield Greenwich Advisors LLC and Fairfield Greenwich (Bermuda) LTD on April 1, 2009.

Markopolos's emails to his team from 2005 until the end of his investigation in April 2008 are taken from documents provided to the U.S. House Committee on Financial Services. His emails cover a range of topics, from the reward money to using options to profit from the collapse of Madoff's scheme.

Markopolos's February 4, 2009, testimony to the U.S. House Committee on Financial Services provided me with a roadmap of his third attempt to have the SEC pursue Bernie Madoff. This extended from his meeting with Mike Garrity, branch chief of the SEC's Boston Regional Office, to his dealings with Meaghan Cheung, branch chief of the New York enforcement division.

I consulted the famous report Markopolos submitted to the SEC, "The World's Largest Hedge Fund Is a Fraud," to understand the twenty-nine red flags that were presented to Mike Garrity and Meaghan Cheung. It is included in Markopolos's House Committee testimony in its multiple stages of revisions and additions. Cheung told me she could not comment on the case.

Roger Madoff's battle with cancer took a terrible toll on the Madoff family, reaching its worst stages during the 2006 SEC investigation. He wrote a book about his struggle, titled *Leukemia for Chickens*. By reading his personal and touching work, we learned about the history of cancer in the Madoff family, Roger's job for *Bloomberg News* in Italy before he worked with his family, and the stem cell transplant Shana gave to save her brother's life. Elaine Solomon was helpful with details on that period as well, as was Eleanor Squillari's *Vanity Fair* article.

The transcript of Bernie Madoff's conversation with Fairfield Greenwich Group's general counsel, Mark McKeefrey, and head of risk management, Amit Vijayvergiya, in December 2005 is found in the administrative complaint filed by the Massachusetts Secretary of the Commonwealth against Fairfield Greenwich Advisors LLC and Fairfield Greenwich (Bermuda) LTD on April 1, 2009.

Madoff lied under oath to SEC investigators about how he operated his investment advisory business on May 19, 2006. The numerous false and misleading statements I quoted are reproduced in the criminal information filed against Bernie Madoff by the Southern District of New York on March 10, 2009.

The SEC's preliminary findings, as well as its final conclusions on the 2006 Madoff case, are found in the SEC Division of Enforcement Case Closing Recommendation, case number NY-07563, released on November 21, 2007.

I was unable to speak with John Wilke of the *Wall Street Journal* about his experience with Harry Markopolos; on May 1, 2009, the 20-year veteran of the *Journal* passed away from pancreatic cancer.

Markopolos describes his decision to end his pursuit of Madoff in April 2008 in his written testimony submitted to the U.S. House Committee on Financial Services and his emails to colleagues.

10. The Fall of Bernie Madoff

President George W. Bush's statement on the looming economic downturn was reported in the January 18, 2008, Associated Press story, "Bush Calls for $145 Billion in Tax Relief." The multibillion-dollar loss by brokerage giant Merrill Lynch was reported in "Merrill Lynch Loss Is Nearly $10 billion," *New York Times*, January 17, 2008. The closing numbers for the Dow Jones Industrial Average were found on www.dowjonesclose.com, a site listing historical closing figures.

The Madoffs' luxurious lifestyle at this time was reconstructed through an analysis of Madoff's American Express bill. The bill for the period December 23, 2007, through January 23, 2008, was released by trustee Irving Picard as Exhibit 25 in a May 5, 2009, affidavit filed with the U.S. Bankruptcy Court. The documents presented to the court attempted to prove that Bernie Madoff was using his business as a personal piggy bank for his friends and family.

Madoff's apartment was described as "Queens High Baroque" by Michael Skakun and Ken Libo. They revisited their time at the Madoff residence in "Sconces and Scrapbooks: A Visit to the Madoffs," *Jewish Daily Forward*, December 18, 2008.

Madoff's seventieth birthday party and the golf tournament at Cabo San Lucas were detailed in "I Just Can't Live with That Camera—It's Not Square: Inside the Bizarre World of £30bn Pyramid Schemester Bernie Madoff," *Daily Mail* (London), January 3, 2009.

During our interview, Stephen Richards discussed the time Madoff rejected Richards's brother-in-law's 401(k) as a new investment. Even in June 2008, Madoff was still turning people down who offered him money.

Internal emails from Fairfield Greenwich struggling to explain Madoff's business model to wary clients are included in exhibits from the Massachusetts Secretary of the Commonwealth's administrative complaint against Fairfield Greenwich Advisors LLC and Fairfield Greenwich (Bermuda) LTD, filed April 1, 2009. Emails in the administrative complaint include Unigestion asking Fairfield whether Madoff was making the trades and holding the assets he was claiming.

Richard Landsberger's email to Amit Vijayvergiya wondering how Madoff was able to find options when liquidity at banks completely dried up is also found in the Massachusetts complaint. The stories of the questionnaire Fairfield asked Madoff to answer before their October 2008 due diligence visit and Jeffrey Tucker's struggling to placate Madoff are both included in the complaint's exhibits.

The chronology and circumstances surrounding JPMorgan Chase's $250 million withdrawal from Fairfield Greenwich was covered in "Madoff's Banker: Where Was JPMorgan Chase?," *Time*, March 25, 2009, and "JPMorgan Exited Madoff-Linked Funds Last Fall," *New York Times*, January 28, 2009. In addition to my questions surrounding Norman Levy's office at Chase headquarters and Levy's depositing checks for millions of dollars, media representatives did not offer further responses to the company's abrupt pull out of Fairfield Greenwich funds.

René-Thierry Magon de la Villehuchet trusting Madoff with even more of his money in the midst of the 2008 credit crunch is attributed to sources who spoke with my assistant, Nathalie Mattheiem, as well as sources interviewed for PBS's May 13, 2009, edition of *Frontline*.

Eleanor Squillari's memories of Bernie in the weeks before the scandal broke were in "Hello, Madoff!," *Vanity Fair*, May 2009. Elaine Solomon was able to confirm many of Squillari's stories and offered a plethora of her own during my multiple conversations with her. Both women were a tremendous help in depicting Bernie's actions, attitude, and stress in the final days before his arrest.

The January 2008 form Madoff filed with the SEC claiming his investment advisory had twenty-three clients and $17.1 billion under management was found on the SEC's Investment Adviser Public Disclosure website. The information is found on a Form ADV, which is an annually renewed document that must be sent to the SEC by investment advisors managing usually more than $25 million. Information about the advisor's business and history of SEC investigations within the previous ten years is included on an ADV.

On a January 21, 2009, edition of *CNBC Reports*, Ken Langone talked about the investment presentation Madoff made to him days before the scam was revealed. Dialogue of what happened in the office is taken from the interview transcript.

Ruth Shapiro remembered how stressed Bernie was when he asked her husband for $250 million in "Palm Beacher Calls Bernie Madoff Arrest 'Knife in the Heart,' " *Palm Beach Daily News*, December 16, 2008.

Ruth Madoff's withdrawals of $15.5 million from her Cohmad account on November 25, 2008, and December 10, 2008, are documented in the Massachusetts Secretary of the Commonwealth's administrative complaint against Cohmad Securities Corporation, February 11, 2009.

I consulted the June 2009 *Vanity Fair* story, "Did the Sons Know?," along with my own interviews with Madoff employees to put together the timeline of events for Madoff's confession—first to Peter, then to his sons, and then to the FBI.

Clive Brown, the driver who revealed the story of Norman Levy's checks, was also the one who drove Bernie, Mark, and Andy to the Madoffs' Lexington Avenue penthouse before Bernie confessed to his sons.

During interviews with me, Elaine Solomon, Clive Brown, and Errol Sibley (Mark and Andy's driver) vividly recounted the firm's 2008 Christmas party, proving extremely helpful for my description of what could have been Bernie's final happy moments before the world knew of his crimes.

FBI Agent Theodore Cacioppi's account of his arrest of Bernie Madoff comes from the Southern District of New York's complaint filed against Madoff on December 11, 2008.

I spoke at length with a spokesman for Andy and Mark Madoff, who offered me a chronology of their experiences during the tumultuous days of December, 2008.

When I spoke with Ira Sorkin shortly after Madoff was sentenced, he described for me how and when he learned Bernie had been arrested: while visiting his granddaughter's nursery school full of children having fun making animal noises.

Errol Sibley was invaluable for his account of December 11, when he witnessed Mark Madoff break down in front of his wife and family help, telling them that his father was a crook.

11. Shattered Lives

Elaine Solomon contributed much of the description of Madoff's office on the day after his arrest. In my interview with her, Solomon also confirmed and elaborated on the stories Eleanor Squillari told in "Hello, Madoff,!" *Vanity Fair*, May 2009, which I also drew on.

Carl Shapiro told the *Palm Beach Daily News* how he heard about Madoff's arrest in "Palm Beacher Calls Bernie Madoff Arrest 'Knife in the Heart,' " December 16, 2008.

Palm Beach accountant Richard Rampell was able to tell me the extent of the Madoff losses suffered by several of his clients ($2 million to $80 million) during our interview.

Yeshiva University's Madoff losses were reported in "Yeshiva Had Losses of $110 Million Linked to Madoff," *Bloomberg*, December 16, 2008.

Mortimer Zuckerman's losses were reported in "Zuckerman Sues Merkin over $40 Million Madoff Loss," *Bloomberg*, April 9, 2009.

Robert Lappin's losses were reported in "Lappin: Justice Has Been Served," *Salem News*, June 30, 2009.

The *New York Times* reported the Ramaz School's Madoff exposure in "Wall St. Fraud Leaves Charities Reeling," December 15, 2008.

Merkin's exposure to Madoff at $2.4 billion has been widely reported, but the figure is also specified in New York Attorney General Andrew Cuomo's complaint against J. Ezra Merkin and Gabriel Capital Corporation, filed April 6, 2009.

Elie Wiesel discussed the losses he and his foundation sustained during a panel discussion arranged by *Portfolio* on February 26, 2009.

An anonymous reader's anti-Jewish comments posted on a *Palm Beach Post* writer's blog come from Jose Lambiet's Page2 Live Blog. I also used Lambiet's blog, among other sources, for details surrounding the altercation between Jerome Fisher and Robert Jaffe, specifically the December 16, 2008, post, "Madoff Investors Face Off in Palm Beach."

During my February 2009 trip to Palm Beach, I spent time at Royal

Pawn and Jewelry interviewing employees about the parade of Madoff victims and the items they sold.

During our interview, Stephen and Fran Richards recounted how they heard about the scam, and Fran described how she broke the news to Bob and Joan Roman. The Friedmans told me of how they cried together in their pajamas when they learned of Madoff's crimes.

The details surrounding the packages Bernie and Ruth sent to friends and family containing millions of dollars in valuable watches and jewelry were found in a brief filed by federal prosecutors on January 7, 2009.

Harry Markopolos's memories of how he heard about Madoff's arrest came from his interview with WBUR Radio on April 21, 2009.

The statement issued by SEC Chairman Christopher Cox admitting the regulator was wrong to ignore Markopolos was sent out in an SEC press release on December 16, 2008.

Losses sustained by Fairfield Greenwich, Banco Santander, and Bank Medici were reported in "Giant Bank in Probe over Ties to Madoff," *Wall Street Journal*, January 13, 2009.

Union Bancaire Privée's losses were reported in "UBP Hires Two Fund Executives in Effort to Rebound from Madoff," *Bloomberg*, July 7, 2009. Fortis's losses were reported in "Madoff's 'Lie' Ensnares Victims from Paris to Tokyo," *Bloomberg*, December 15, 2008.

The majority of the reporting, description, and translation of the passages about the suicide of René-Thierry Magon de la Villehuchet was conducted by a talented freelance reporter who worked with me on this book, Nathalie Mattheiem.

De la Villehuchet told a friend he was lucky to invest with Madoff, according to "Le Suicide pour l'Honneur de Thierry de la Villehuchet," *Paris Match*, December 30, 2008. *Paris Match* also outlines a timeline of the events surrounding his death. Nathalie Mattheiem interviewed Guy Gurney for his recollection of how de la Villehuchet struggled to cope with the Madoff disaster.

The story of William Foxton's suicide was reported in "Army Major Kills Himself over Bernard Madoff Fraud Debts," *Telegraph* (London), February 13, 2009.

Clark Landis, the financial trader who stood outside Madoff's building with the sign urging him to jump, was on the cover of the *New York Post* on January 15, 2008, and featured in the article "Frightened Bernie Now Bulletproof."

Bernie Madoff's plea for Ruth to keep their penthouse apartment and $70 million in assets is found in a footnote in a federal court filing made by prosecutors on March 2, 2009.

Jerry Horowitz's son, Irwin, wrote about his father's anguish after news of the scam broke, and of his father's death the day Madoff pled guilty, on the *New West Community Blog* post, "My Father Died Today," March 12, 2009.

Allegations that Sonja Kohn received $40 million in kickbacks from Bernie Madoff to funnel clients his way were reported in "British Study Madoff Payments to Austrian Banker," *New York Times*, July 3, 2009.

Court-appointed trustee Irving Picard alleged in a May 12 complaint against investor Jeffrey Picower that Picower raked in at least $5 billion in Madoff profits between December 1995 and December 2008. Because the returns were implausibly high, the lawsuit asserts that Picower should have known the profits were generated through fraudulent activity.

My final pages on Bernie Madoff's dramatic sentencing come from the observations I made in the courtroom that morning, when Madoff learned the price he would pay for his staggering betrayal.

Index